Interactive Mathematics Program®

I M P

YEAR **4**

Integrated High School Mathematics

High Dive

Dan Fendel and Diane Resek
with
Lynne Alper and Sherry Fraser

KEY CURRICULUM PRESS
Innovators in Mathematics Education

This material is based upon work supported
by the National Science Foundation
under award number ESI-9255262.
Any opinions, findings, and conclusions
or recommendations expressed in this
publication are those of the authors
and do not necessarily reflect the views
of the National Science Foundation.

Key Curriculum Press
1150 65th Street
Emeryville, California 94608
510-595-7000
editorial@keypress.com
http://www.keypress.com

10 9 8 7 6 5 4 3 2 1   03 02 01 00 99
ISBN 1-55953-345-5

Printed in the United States of America

Project Editor
Casey FitzSimons

Project Administrator
Jeff Gammon

Additional Editorial Development
Masha Albrecht, Mary Jo Cittadino

Art Developer
Ellen Silva

Production Editors
Caroline Ayres, Kristin Ferraioli

Project Assistants
Stefanie Liebman, Beck Finley

Copyeditor
Thomas L. Briggs

Cover and Interior Design
Terry Lockman
Lumina Designworks

Production and Manufacturing Manager
Diana Jean Parks

Production Coordinator
Laurel Roth Patton

Art Editor
Kelly Murphy

Photo Researcher
Laura Murray

Technical Graphics
Tom Webster, Lineworks, Inc.

Illustration
Alan Dubinsky, Tom Fowler, Nikki Middendorf, Evangelia Philippidis,
Paul Rogers, Sara Swan, Martha Weston, April Goodman Willy,
Amy Young

Publisher
Steven Rasmussen

Editorial Director
John Bergez

MATHEMATICS REVIEW
Lynn Arthur Steen, St. Olaf College,
   Northfield, Minnesota

MULTICULTURAL REVIEWS
Genevieve Lau, Ph.D., Skyline College,
   San Bruno, California
Luis Ortiz-Franco, Ph.D., Chapman University,
   Orange, California
Marilyn E. Strutchens, Ph.D., University of Maryland,
   College Park, Maryland

TEACHER REVIEWS
Kathy Anderson, Aptos, California
Dani H. Brutlag, Mill Valley, California
Robert E. Callis, Oxnard, California
Susan Schreibman Ford, Stockton, California
Mary L. Hogan, Arlington, Massachusetts
Jane M. Kostik, Minneapolis, Minnesota
Brian R. Lawler, Carson, California
Brent McClain, Portland, Oregon
Michelle Novotny, Aurora, Colorado
Barbara Schallau, San Jose, California
James Short, Oxnard, California
Kathleen H. Spivack, New Haven, Connecticut
Linda Steiner, Escondido, California
Marsha Vihon, Chicago, Illinois
Edward F. Wolff, Glenside, Pennsylvania

Acknowledgments

Many people have contributed to the development of the IMP™ curriculum, including the hundreds of teachers and many thousands of students who used preliminary versions of the materials. Of course, there is no way to thank all of them individually, but the IMP directors want to give some special acknowledgments.

We want to give extraordinary thanks to these people who played unique roles in the development of the curriculum.

• **Matt Bremer** did the initial revision of every unit after its pilot testing. Each unit of the curriculum also underwent extensive focus-group reexamination after being taught for several years, and Matt rewrote many units following the focus groups. He has read every word of everyone else's revisions as well and has contributed tremendous insight through his understanding of high school students and the high school classroom.

• **Mary Jo Cittadino** became a high school student once again during the piloting of the curriculum, attending class daily and doing all the class activities, homework, and POWs. Because of this experience, her contributions to focus groups had a unique perspective. This is a good place to thank her also for her contributions to IMP as Network Coordinator for California. In that capacity, she visited many IMP classrooms and answered thousands of questions from parents, teachers, and administrators.

• **Bill Finzer** was one of the original directors of IMP before going on to different pastures. Though he was not directly involved in the writing of Year 4, he made important contributions to the development of *High Dive,* and his ideas about curriculum are visible throughout the program.

• **Lori Green** took a leave from the classroom as a regular teacher after the 1989–1990 school year and became a traveling resource for IMP classroom teachers. In that role, she has seen more classes using the curriculum than we can count. She has compiled many of the insights from her classroom observations in the *Teaching Handbook for the Interactive Mathematics Program*®.

• **Celia Stevenson** developed the charming and witty graphics that graced the prepublication versions of all the IMP units.

Several people played particular roles in the development of this unit, *High Dive.*

• Dean Ballard, Matt Bremer, Donna Gaarder, Steve Jenkins, Dan Johnson, Jean Klanica, Barbara Schallau, and Adrienne Yank helped us create the version of *High Dive* that was pilot tested during 1993–1994. They not only taught the unit in their classrooms that year, but they also read and commented on early drafts, tested almost all the activities during workshops that preceded the teaching, and then came back after teaching the unit with insights that contributed to the initial revision.

- Susan Ford, Dan Johnson, and Jim Short joined Matt Bremer for the focus group on *High Dive* in November 1996. Their contributions built on several years of IMP teaching, including at least two years teaching this unit, and their work led to the development of the last field-test version of the unit.

- Matt Bremer, Steve Hansen, Mary Hunter, and Caran Resciniti field tested the post-focus-group version of *High Dive* during 1997–1998. Matt and Caran met with us to share their experiences when the teaching of the unit was finished. Their feedback helped shape the final version that now appears.

In creating this program, we needed help in many areas other than writing curriculum and giving support to teachers.

The National Science Foundation (NSF) has been the primary sponsor of the Interactive Mathematics Program. We want to thank NSF for its ongoing support, and we especially want to extend our personal thanks to Dr. Margaret Cozzens, who was Director of NSF's Division of Elementary, Secondary, and Informal Education during IMP's development period, for her encouragement and her faith in our efforts.

We also want to acknowledge here the initial support for curriculum development from the California Postsecondary Education Commission and the San Francisco Foundation, and the major support for dissemination from the Noyce Foundation and the David and Lucile Packard Foundation.

Keeping all of our work going required the help of a first-rate office staff. This group of talented and hard-working individuals worked tirelessly on many tasks, such as sending out units, keeping the books balanced, helping us get our message out to the public, and handling communications with schools, teachers, and administrators. We greatly appreciate their dedication.

- Barbara Ford—Secretary

- Tony Gillies—Project Manager

- Marianne Smith—Communications Manager

- Linda Witnov—Outreach Coordinator

We want to thank Dr. Norman Webb of the Wisconsin Center for Education Research for his leadership in our evaluation program, and our Evaluation Advisory Board, whose expertise was so valuable in that aspect of our work.

- David Clarke, University of Melbourne

- Robert Davis, Rutgers University

- George Hein, Lesley College

- Mark St. John, Inverness Research Associates

IMP National Advisory Board

Finally, we want to thank Steve Rasmussen, President of Key Curriculum Press, John Bergez, Key's Executive Editor for the IMP curriculum, Casey FitzSimons, Project Editor, and the many others at Key whose work turned our ideas and words into published form.

Dan Fendel   Diane Resek   Lynne Alper   Sherry Fraser

Foreword

"I hated math" is an often-heard phrase that reflects an unfortunate but almost socially acceptable adult prejudice. One hears it from TV announcers, politicians, and even football coaches. I'll bet they didn't start their education feeling that way, however. According to the Third International Mathematics and Science Study, American fourth graders score near the top of their international peers in science and math. Surely mathphobia hasn't broken out by that grade level. By twelfth grade, however, students in the United States score among the lowest of the 21 participating nations in both mathematics and science general knowledge. Even our advanced math students—the ones we like to think are the best in the world—score at the very bottom when compared to advanced students in other countries. What happened? Is there something different about our students? Not likely. Is there an opportunity for improvement in our curriculum? You bet.

Traditional mathematics teaching continues to cover more repetitive and less challenging material. For the majority of students, rote memorization, if not too difficult, is certainly an unenlightening chore. The learning that does result tends to be fragile. There is little time to gain deep knowledge before the next subject has to be covered. American eighth-grade textbooks cover five times as many subjects in much less depth than student materials found in Japan. Because there is no focus on helping students discover fundamental mathematical truths, traditional mathematics education in the United States fails to prepare students to apply knowledge to problems that are slightly different and to situations not seen before.

As an engineering director in the aerospace industry, I'm concerned about the shrinking supply of talented workers in jobs that require strong math and science skills. In an internationally competitive marketplace, we desperately need employees who have not only advanced academic skills, but also the capability to discover new, more cost-effective ways of doing business. They need to design with cost as an independent variable. They need to perform system trades that not only examine the traditional solutions, but explore new solutions through lateral, "out of the box" thinking. They need to work in teams to solve the most difficult problems and present their ideas effectively to others.

Programs like IMP foster these skills and fulfill our need as employers to work with educators to strengthen the curriculum, making it more substantive and challenging. I can attest to the value of IMP because, as the father of a student who has completed four years of the program, I've discovered that something *different* is going on here. My son is given problems around a theme, each one a little harder than the one before. This is not much different from the way I was taught. What is different is that he is not given the basic math concept ahead of time, nor is he shown how to solve upcoming problems by following the rule. By attacking progressively harder problems in many different ways, he often learns the basic mathematical concepts through discovery. He is taught to think for himself. He says that the process "makes you feel like you are actually solving the problem, not just repeating what the teacher says."

This process of encouraging discovery lies at the heart of IMP. Discovery is not fragile learning; it is powerful learning. My son thinks it can be fun, even if he won't admit it to other students.

I have another window on IMP as well. As the husband of a teacher who helped to pioneer the use of IMP in her district, I've learned that teaching IMP is a lot more than letting the students do their own thing. Lessons are carefully chosen to facilitate the discovery process. Points are given for finding the correct answer, and points are given for carefully showing all work, which is as it should be. Because the curriculum encourages different ways of solving a problem, my wife spends more than the typical amount of time teachers spend in reading and understanding students' efforts. The extra time doesn't seem to burden her, however. I think she thinks it's fun. She even gets excited when she sees that the focus on communicating and presenting solutions is measurably improving her students' English skills.

Let me conclude with a word of encouragement to all of you who are using this book. I congratulate you for your hard work and high standards in getting to this, the fourth and final year of IMP. IMP students have performed well in SAT scores against their peers in traditional programs. Colleges and universities accept IMP as a college preparatory mathematics sequence. I know that your efforts will pay off, and I encourage you to take charge of your future by pursuing advanced math and science skills. Even if you don't become an aerospace engineer or computer programmer, this country needs people who think logically and critically, and who are well prepared to solve the issues yet to be discovered.

Larry Gilliam
Scotts Valley, California

Larry Gilliam is a parent of two IMP students and works as the chief test engineer for Lockheed Martin Missiles & Space in Sunnyvale, California.

The Interactive Mathematics Program

What is the Interactive Mathematics Program?

The Interactive Mathematics Program (IMP) is a growing collaboration of mathematicians, teacher-educators, and teachers who have been working together since 1989 on both curriculum development and professional development for teachers.

What is the IMP curriculum?

IMP has created a four-year program of problem-based mathematics that replaces the traditional Algebra I–Geometry–Algebra II/Trigonometry–Precalculus sequence and that is designed to exemplify the curriculum reform called for in the *Curriculum and Evaluation Standards* of the National Council of Teachers of Mathematics (NCTM).

The IMP curriculum integrates traditional material with additional topics recommended by the NCTM *Standards,* such as statistics, probability, curve fitting, and matrix algebra. Although every IMP unit has a specific mathematical focus, most units are structured around a central problem and bring in other topics as needed to solve that problem, rather than narrowly restricting the mathematical content. Ideas that are developed in one unit are generally revisited and deepened in one or more later units.

For which students is the IMP curriculum intended?

The IMP curriculum is for all students. One of IMP's goals is to make the learning of a core mathematics curriculum accessible to everyone. Toward that end, we have designed the program for use with heterogeneous classes. We provide you with a varied collection of supplemental problems to give you the flexibility to meet individual student needs.

Dan Johnson and Susan Ford use a diagram as they consider a diver's initial velocity at the moment of release from the Ferris wheel's platform.

How is the IMP classroom different?

When you use the IMP curriculum, your role changes from "imparter of knowledge" to observer and facilitator. You ask challenging questions. You do not give all the answers; rather, you prod students to do their own thinking, to make generalizations, and to go beyond the immediate problem by asking themselves "What if?" The IMP curriculum gives students many opportunities to write about their mathematical thinking, to reflect on what they have done, and to make oral presentations to one another about their work. In IMP, your assessment of students becomes integrated with learning, and you evaluate students according to a variety of criteria, including class participation, daily homework assignments, Problems of the Week, portfolios, and unit assessments. The *Teaching Handbook for the Interactive Mathematics Program* provides many practical suggestions on how to get the best possible results using this curriculum in *your* classroom.

What is in Year 4 of the IMP curriculum?

Year 4 of the IMP curriculum contains five units.

High Dive

The central problem of this unit involves a circus act in which a diver is dropped from a turning Ferris wheel into a tub of water carried by a moving cart. The students' task is to determine when the diver should be released from the Ferris wheel in order to land in the moving tub of water. In analyzing this problem, students extend right-triangle trigonometric functions to the circular functions, study the physics of falling objects (including separating the diver's motion into its vertical and horizontal components), and develop an algebraic expression for the time of the diver's fall in terms of his position. Along the way, students are introduced to several additional trigonometric concepts, such as polar coordinates, inverse trigonometric functions, and the Pythagorean identity.

As the Cube Turns

This unit opens with an overhead display, generated by a program on a graphing calculator. The two-dimensional display depicts the rotation of a cube in three-dimensional space. Students' central task in the unit is to learn how to write such a program, though the real focus is on the mathematics behind the program.

Students study the fundamental geometric transformations—translations, rotations, and reflections—in both two and three dimensions and express them in terms of coordinates. The study of these transformations also provides a new setting for students to work with matrices, which they previously studied in connection with systems of linear equations (in the Year 3 unit *Meadows or Malls?*). Another complex component of students' work is the analysis of how to represent a three-dimensional object on a two-dimensional screen. As a concluding project, students work in pairs to program an animated graphic display of their own design.

Know How

This unit is designed to prepare students to find out independently about mathematical content they either have not learned or have forgotten. Most will need this skill in later education as well as in their adult work lives. Students are given experiences of learning through reading traditional textbooks and interviewing other people. The content explored this way includes radian measure, ellipses, proof of the quadratic formula, the laws of sines and cosines, and complex numbers.

The World of Functions

This unit builds on students' extensive previous work with functions. Students explore basic families of functions in terms of various ways they

can be represented—as tables, as graphs, as algebraic expressions, and as models for real-world situations. Students also use functions to explore a variety of problem situations and discover that finding an appropriate function to use as a model sometimes involves recognizing a pattern in the data and other times requires insight into the situation itself. In the last portion of the unit, students explore ways of combining and transforming functions.

The Pollster's Dilemma

The central limit theorem is the cornerstone of this unit in which students look at the process of sampling, with a special focus on how the size of the sample affects variation in poll results. The opening problem concerns an election poll, which shows 53% of the voters favoring a particular candidate.

Students investigate this question: How confident should the candidate be about her lead, based on this poll? By analyzing specific cases, students see that the results from a set of polls of a given size are approximately normally distributed. They are given the statement of the central limit theorem, which confirms this observation. Building on work in earlier units, students learn how to use normal distributions and standard deviations to find confidence intervals and see how concepts such as margin of error are used in reporting polling results. Students finish the unit by working in pairs on a sampling project for a question of their own.

How do the four years of the IMP curriculum fit together?

The four years of the IMP curriculum form an integrated sequence through which students can learn the mathematics they will need both for further education and on the job. Although the organization of the IMP curriculum is very different from the traditional Algebra I–Geometry–Algebra II/Trigonometry–Precalculus sequence, the important mathematical ideas are all there.

Here are some examples of how both traditional concepts and topics new to the high school curriculum are developed.

Solving equations

In Year 1 of the IMP curriculum, students develop an intuitive foundation of algebraic thinking, including the use of variables, which they build on throughout the program. In the Year 2 unit *Solve It!,* students use the concept of equivalent equations to see how to solve any linear equation in a single variable. In *Cookies* (Year 2) and *Meadows or Malls?* (Year 3), they solve pairs of linear equations in two or more variables, using both algebraic and geometric methods, and see how to use matrices and the technology of graphing calculators to solve such systems. In *Fireworks* (Year 3), they explore a variety of methods for solving quadratic equations, including graphing and completing the square. In *High Dive* and *Know How* (Year 4), students use the quadratic formula to solve an equation that arises from the study of falling objects, and they prove the general formula using ideas from *Fireworks.*

Geometry in One, Two, and Three Dimensions

Measurement, including area and volume, is one of the fundamental topics in geometry. In Year 1, students use angle and line measurement and their relationship to similarity to explore ideas about lengths in the unit *Shadows.* In *Do Bees*

Build It Best? (Year 2), students discover and prove the Pythagorean theorem and develop important ideas about area, volume, and surface area. Students combine these ideas with their understanding of similarity to see why the hexagonal prism of the bees' honeycomb design is the most efficient regular prism possible. In the Year 3 unit *Orchard Hideout,* students examine why the special number π appears in the formulas for both area and circumference of circles. They use these formulas, together with principles of coordinate geometry, to predict how long it will take for the center of an orchard to become a "hideout." Later in Year 3, in *Meadows or Malls?,* students extend ideas of coordinate graphing from two dimensions to three, and apply key ideas about three-dimensional graphs to solve a land-use problem. Work in three dimensions continues in Year 4, especially in *As the Cube Turns,* in which students examine how to represent a three-dimensional figure mathematically using a two-dimensional diagram.

Trigonometric functions

In traditional programs, the trigonometric functions are introduced in the eleventh or twelfth grade. In the IMP curriculum, students begin working with trigonometry in Year 1 in the unit *Shadows* and use right-triangle trigonometry in several units in Years 2 and 3, including the unit *Do Bees Build It Best?* In the Year 4 unit *High Dive,* they extend trigonometry from right triangles to circular functions in the context of a circus act in which a performer is dropped from a Ferris wheel into a moving tub of water.

Standard deviation and the binomial distribution

Standard deviation and the binomial distribution are major tools in the study of probability and statistics. The Year 1 unit *The Game of Pig* gets students started by building a firm understanding of concepts of probability and the phenomenon of experimental variation. Later in Year 1 (in *The Pit and the Pendulum*), they use standard deviation to see that the period of a pendulum is determined primarily by its length. In Year 2, students compare standard deviation with the chi-square test in examining whether the difference between two sets of data is statistically significant. In *Pennant Fever* (Year 3), students use the binomial distribution to evaluate a team's chances of winning the baseball championship, and in *The Pollster's Dilemma* (Year 4), students tie many of these ideas together in the central limit theorem, seeing how the margin of error and the level of certainty for an election poll depend on the size of the sample.

Does the program work?

The IMP curriculum has been thoroughly field-tested and enthusiastically received by hundreds of classroom teachers around the country. Their enthusiasm is based on the success they have seen in their own classrooms with their own students. For instance, IMP teacher Dennis Cavaillé says, "For the first time in my teaching career, I see lots of students excited about solving math problems inside *and* outside of class."

These informal observations are backed up by more formal evaluations. Dr. Norman Webb of the Wisconsin Center for Education Research has done several studies comparing the performance of students using the IMP curriculum with the performance of students in traditional programs. For instance, he has found that IMP students do as well as students in traditional mathematics classes on standardized tests such as the SAT. This is especially significant because IMP students spend about twenty-five percent of their time studying topics, such as statistics, not covered on these tests. To measure IMP students' achievement in these other areas, Dr. Webb conducted three separate studies involving students at different grade levels and in different locations. The three tests used in these studies involved statistics, quantitative reasoning, and general problem solving. In all three cases, the IMP students outperformed their counterparts in traditional programs by a statistically significant margin, even though the two groups began with equivalent scores on eighth grade standardized tests.

But one of our proudest achievements is that IMP students are excited about mathematics, as shown by Dr. Webb's finding that they take more mathematics courses in high school than their counterparts in traditional programs. We think this is because they see that mathematics can be relevant to their own lives. If so, then the program works.

Dan Fendel

Diane Resek

Lynne Alper

Sherry Fraser

WELCOME!

Note to Students

This textbook represents the last year of a four-year program of mathematics learning and investigation. As in the first three years, the program is organized around interesting, complex problems, and the concepts you learn grow out of what you'll need to solve those problems.

If you studied IMP Year 1, 2, or 3

If you studied IMP Year 1, 2, or 3, then you know the excitement of problem-based mathematical study. The Year 4 program extends and expands the challenges that you worked with previously. For instance:

• In Year 1, you began developing a foundation for working with variables. In Year 2, you learned how to solve linear equations algebraically, and in Year 3, you worked with quadratic equations. In Year 4, you'll solve a quadratic equation as part of the process of finding out when a diver should be dropped from a Ferris wheel in order to land in a moving tub of water.

• In Year 1, you used the normal distribution to help predict the period of a 30-foot pendulum. In Year 2, you learned about the chi-square statistic to understand statistical comparisons of populations, and in Year 3, you learned about the binomial distribution. In Year 4, you'll use the context of election polls to see the connection between the binomial distribution and the normal distribution, and you'll use ideas such as margin of error and confidence level to study how sample size affects poll reliability.

You'll also use ideas from geometry, trigonometry, and matrix algebra to develop a calculator program that shows a cube rotating in space, you'll prove the quadratic formula as part of a unit on ways to learn mathematics on your own, and you'll synthesize your IMP experience with functions by examining a variety of methods for creating functions that fit specific real-world problems.

These pages in the student book welcome students to the program.

Welcome!

If you didn't study IMP Year 1, 2, or 3

If this is your first experience with the Interactive Mathematics Program (IMP), you can rely on your classmates and your teacher to fill in what you've missed. Meanwhile, here are some things you should know about the program, how it was developed, and how it is organized.

The Interactive Mathematics Program is the product of a collaboration of teachers, teacher-educators, and mathematicians who have been working together since 1989 to reform the way high school mathematics is taught. About one hundred thousand students and five hundred teachers used these materials before they were published. Their experiences, reactions, and ideas have been incorporated into this final version.

Our goal is to give you the mathematics you need in order to succeed in this changing world. We want to present mathematics to you in a manner that reflects how mathematics is used and that reflects the different ways people work and learn together. Through this perspective on mathematics, you will be prepared both for continued study of mathematics in college and for the world of work.

This book contains the various assignments that will be your work during Year 4 of the program. As you will see, these problems require ideas from many branches of mathematics, including algebra, geometry, probability, graphing, statistics, and trigonometry. Rather than present each of these areas separately, we have integrated them and presented them in meaningful contexts, so you will see how they relate to each other and to our world.

Each unit in this four-year program has a central problem or theme, and focuses on several major mathematical ideas. Within each unit, the material is organized for teaching purposes into "days," with a homework assignment for each day. (Your class may not follow this schedule exactly, especially if it doesn't meet every day.)

At the end of the main material for each unit, you will find a set of supplementary problems. These problems provide you with additional opportunities to work with ideas from the unit, either to strengthen your understanding of the core material or to explore new ideas related to the unit.

Although the IMP program is not organized into courses called "Algebra," "Geometry," and so on, you will be learning all the essential mathematical concepts that are part of those traditional courses. You will also be learning concepts from branches of mathematics—especially statistics and probability—that are not part of a traditional high school program.

To accomplish your goals, you will have to be an active learner, because the book does not teach directly. Your role as a mathematics student will be to experiment, to investigate, to ask questions, to make and test conjectures, and to reflect, and then to communicate your ideas and conclusions both orally and in writing. You will do some of your work in collaboration with fellow students, just as users of mathematics in the real world often work in teams. At other times, you will be working on your own.

We hope you will enjoy the challenge of this new way of learning mathematics and will see mathematics in a new light.

Dan Fendel   Diane Resek

Lynne Alper   Sherry Fraser

Finding What You Need

We designed this guide to help you find what you need amid all the information it provides. Each of the following components has a special treatment in the layout of the guide.

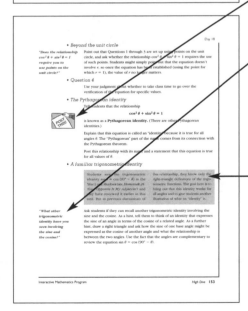

Synopsis of the Day: The key idea or activity for each day is summarized in a brief sentence or two.

Mathematical Topics: Mathematical issues for the day are presented in a bulleted list.

Outline of the Day: Under the *In Class* heading, the outline summarizes the activities for the day, which are keyed to numbered headings in the discussion. Daily homework assignments and Problems of the Week are listed under the *At Home* heading.

Special Materials Needed: Special items needed in the classroom for each day are bulleted here.

Discuss With Your Colleagues: This section highlights topics that you may want to discuss with your peers.

Post This: The *Post This* icon indicates items that you may want to display in the classroom.

Suggested Questions: These are specific questions that you might ask during an activity or discussion to promote student insight or to determine whether students understand an idea. The appropriateness of these questions generally depends on what students have already developed or presented on their own.

Asides: These are ideas outside the main thrust of a discussion. They include background information, refinements or subtle points that may only be of interest to some students, ways to help fill in gaps in understanding the main ideas, and suggestions about when to bring in a particular concept.

Icons for Student Written Products

Single group report

Individual reports

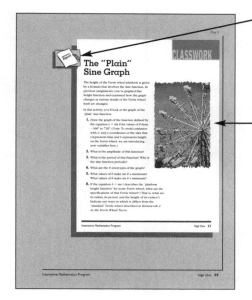

Icons for Student Written Products: For each group activity, there is an icon suggesting a single group report, individual reports, or no report at all. If graphs are included, the icon indicates this as well. (The graph icons do not appear in every unit.)

Embedded Student Pages: The teacher guide contains reduced-size copies of the pages from the student book, including the "transition pages" that appear occasionally within each unit to summarize each portion of the unit and to prepare students for what is coming. The reduced-size classwork and homework assignments follow the teacher notes for the day on which the activity is begun. Having all of these student pages in the teacher's guide is a helpful way for you to see things from the students' perspective.

Additional Information

Here is a brief outline of other tools we have included to assist you and make both the teaching and the learning experience more rewarding.

Glossary: This section, which is found at the back of the book, gives the definitions of important terms for all of Year 4 for easy reference. The same glossary appears in the student book.

Appendix A: Supplemental Problems: This appendix contains a variety of interesting additional activities for the unit, for teachers who would like to supplement material found in the regular classroom problems. These additional activities are of two types—*reinforcements,* which help increase student understanding of concepts that are central to the unit, and *extensions,* which allow students to explore ideas beyond the basic unit.

Appendix B: Blackline Masters: For each unit, this appendix contains materials you can reproduce that are not available in the student book and that will be helpful to teacher and student alike. They include the end-of-unit assessments as well as such items as diagrams from which you can make transparencies. Semester assessments for Year 4 are included in *As the Cube Turns* (for first semester) and *The Pollster's Dilemma* (for second semester).

Single group graph

Individual graphs

No report at all

Year 4 IMP Units

High Dive (in this book)

As the Cube Turns

Know How

The World of Functions

The Pollster's Dilemma

CONTENTS

High Dive Overview

Summary of the Unit

The central problem of this unit concerns a circus act in which a diver is dropped from a turning Ferris wheel into a tub of water carried by a moving cart. The basic problem is to determine when his fall should begin in order for him to land in the water. Students begin by looking at the diver's height off the ground, analyzing different cases in terms of the angle through which the Ferris wheel has turned, and seeing that the analysis is slightly different from one quadrant to another.

Students then develop a formula based on right-triangle trigonometry that works when the diver is in the first quadrant. They use this formula as a clue for how to extend the sine function from the familiar right-triangle context so that it is defined for all angles. The development of the general definition of the sine function involves a blend of considerations, including the physical situation of the Ferris wheel, a graph of the diver's height, and a coordinate model of the wheel.

The cosine function is developed similarly and reinforces the elegance and power of definitions involving the coordinate system. In particular, students see that these definitions eliminate the need for a quadrant-by-quadrant analysis, incorporating the issue of sign quite nicely.

Students also learn about the graphs of the sine and cosine functions. In particular, they see how the graphs of the functions describing the diver's position change as various parameters such as the radius of the Ferris wheel and the period of its motion change.

The physics and mathematics of falling objects compose another major strand of the unit, based on the principle that falling objects have constant acceleration. Inititally, students are told to assume for simplicity that the diver is falling as if from rest. They develop an expression for the height of an object falling from rest in terms of its time in the air, which is then used to determine the duration of the diver's fall. In the process, they review ideas about instantaneous and average speed and interpret these concepts graphically.

This work on falling objects is then combined with the analysis of the diver's position on the Ferris wheel and with information about the speed of the cart. Students synthesize these parts of the problem to develop a complex equation involving the amount of time the diver should stay on the Ferris wheel before being dropped. Then they solve this equation graphically (because it is too complex to be susceptible to algebraic manipulation) to find a preliminary solution to the unit problem.

After solving this equation, students remove the assumption that the diver is falling as if from rest and look at the initial velocity given to the diver through the turning of the Ferris wheel. They first examine how an initial vertical component of velocity, either upward or downward, changes the diver's falling time. They see that finding the diver's falling time requires solving a quadratic equation, which leads to an excursion into the quadratic formula.

Students also must grapple with the task of determining both the vertical and horizontal components of the diver's initial velocity, and they must determine how the horizontal component of his initial velocity affects where he lands. The issues related to finding the separate components are dealt with in a series of paired problems, with one problem in each pair involving the Ferris wheel situation and the other set in some other context. Students develop a general expression, based on the physical context, for the time it takes for a falling object to reach the ground, in terms of its initial height and velocity.

Finally, students return to the circus-act problem in its full complexity. They combine their formula for falling time with expressions for the vertical and horizontal components of the diver's velocity. This leads to a very complex expression for the diver's position when he is about to land, in terms of the time of his release from the Ferris wheel. Comparison of this with the position of the moving cart leads to an equation that will solve the problem. This equation is also solved graphically.

Along the road to solving the main unit problem, students pursue various digressions about trigonometric functions, including the Pythagorean identity and other trigonometric identities, the idea of polar coordinates, and inverse sine and cosine functions and their principal values.

Note: Some teachers might prefer to skip the more complex version of the unit problem. Details on how to limit the unit this way are included on Day 16.

On a day-by-day basis, the unit looks roughly like this.

- Days 1–2: Introducing the unit problem

- Days 3–7: Analyzing the diver's height and developing the sine function (with POW presentations on Day 7)

- Days 8–10: Studying constant acceleration and developing a formula for the height of an object falling from rest

- Days 11–13: Analyzing the diver's horizontal position and developing the cosine function

- Days 14–16: Solving the simplified version of the unit problem

- Days 17–20: Continuing to work with trigonometric functions, including polar coordinates (with POW presentations on Day 20)

- Days 21–25: Analyzing the height and falling time for falling objects with nonzero initial vertical velocity (with a digression on the quadratic formula on Day 23)

- Days 26–30: Separating velocity into vertical and horizontal components, and analyzing these components for the diver in the unit problem

- Days 31–34: Solving the unit problem, portfolios, unit assessments, and summing up

Concepts and Skills

Here is a summary of the main concepts and skills that students will encounter and practice in this unit.

Trigonometry

- Extending the trigonometric functions to all angles

- Recognizing the importance of similarity in the definitions of the trigonometric functions

- Graphing the trigonometric functions and variations on those functions

- Defining the inverse trigonometric function and principal values

- Discovering and explaining the Pythagorean identity, $\sin^2 \theta + \cos^2 \theta = 1$, and other trigonometric identities

- Defining polar coordinates and developing expressions for rectangular coordinates in terms of polar coordinates

Physics

- Developing quadratic expressions for the height of free-falling objects, based on the principle of constant acceleration

- Recognizing that a person falling from a moving object will fall differently from someone falling from a stationary object

- Expressing velocity in terms of vertical and horizontal components

- Studying the motion of falling objects when the vertical and horizontal components of the initial velocity are both nonzero

Quadratic equations

- Recognizing the importance of quadratic equations in studying falling objects

- Learning how to use the quadratic formula to solve quadratic equations

- Finding a general solution for the falling time of objects with an initial vertical velocity

The two Problems of the Week involve mathematical induction and recursive relationships.

Materials

You will need to provide these materials during the course of the unit (in addition to standard equipment and materials such as graphing calculators, transparencies, graph paper, chart paper, and marking pens).

- Paper plates, pipe cleaners, and toy cars for building models of the problem (Day 1)

- Coins or discs for introducing *POW 1: The Tower of Hanoi* (Day 1)

You may wish to seek out materials from your physics department to demonstrate or explain physics principles for Days 21 and 26. You will also need to make copies or transparencies of materials in Appendix B.

Students need to provide scientific calculators and graph paper for use at home.

Grading

The IMP *Teaching Handbook* contains general guidelines about how to grade students in an IMP class. You will probably want to check daily that students have done their homework and include regularity of homework completion as part of students' grades. Your grading scheme will probably also include Problems of the Week, the unit portfolio, and the end-of-unit assessments.

Because you will not be able to read thoroughly every assignment that students turn in, you will need to select certain assignments to read carefully and to base grades on. Here are some suggestions.

- *Homework 2: As the Ferris Wheel Turns*

- *Testing the Definition* (Days 4–5)

- *Homework 7: More Beach Adventures*

- *Homework 10: A Practice Jump*

- *Moving Cart, Turning Ferris Wheel* (Days 14–16)

- *Three O'Clock Drop* (Days 24–25)

- *Homework 27: Racing the River*

- *"High Dive" Concluded* (Days 31–32)

If you want to base your grading on more tasks, there are many other homework assignments, class activities, and oral presentations you can use.

Interactive Mathematics Program®

Integrated High School Mathematics

I M P

YEAR 4

High Dive

DAYS 1-2

Going to the Circus

This page in the student book introduces Days 1 and 2.

Maribel DeLoa, Vivian Barajas, and Caroline Moo build a physical model as a first step toward solving the unit problem.

The central problem of this unit involves a circus act in which a diver falls from a turning Ferris wheel into a tub of water carried by a moving cart. The problem involves various kinds of motion, and you will need to learn quite a bit of mathematics before the unit is over in order to solve this problem.

The Ferris Wheel

Mathematical Topics

- Studying a complex situation involving various kinds of motion

---

Outline of the Day

In Class

1. Form new groups

2. *High Dive*

- Students examine the setup in the unit problem and pose questions concerning the situation
- The activity will be discussed on Day 2

3. Refer students to *The Standard POW Write-up*

4. Introduce *POW 1: The Tower of Hanoi*

At Home

Homework 1: The Ferris Wheel

POW 1: The Tower of Hanoi
(due on Day 7)

Special Materials Needed

- Items such as paper plates, pipe cleaners, and toy cars for building models of the problem
- Coins or discs for introducing *POW 1: The Tower of Hanoi*
- A transparency of a "clock diagram" (see Appendix B)

---

1. Forming Groups

At the beginning of the unit, group students as described in the IMP *Teaching Handbook,* recording the members of each group and the suit for each student. We recommend that you create new groups again on Day 11 and on Day 21.

2. *High Dive*

You may want to have one or more students read the introduction to the unit problem out loud (only the section entitled "The Circus Act"). Then provide materials to each group so that students can make their own physical model of the problem (as in Question 1). You might provide paper plates and pipe cleaners for the Ferris wheel itself and a toy car to represent the moving cart. Circulate among the groups to see that everyone has the right idea. (You may want to keep one or more of these models on hand for demonstration purposes during the rest of the unit.)

The cart and Ferris wheel are set up similarly to what is shown in the diagram below, with the cart passing in front of the Ferris wheel. The platform, which is not shown in this diagram, is pointed "forward" (out from the page), perpendicular to the plane in which the Ferris wheel turns. The front end of the platform is directly above the path of the cart.

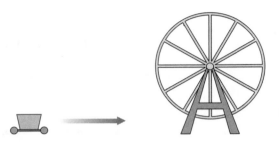

Once everyone has a clear picture of the situation, each group should work on compiling a list of questions as indicated in Question 2, which will be discussed tomorrow.

- *Optional: Pre-Day 1 activity*

  If your school requires a great deal of administrative work on the first day of class, you might consider reading the activity and brainstorming materials that students could use to build their Ferris wheel model. They could then bring these to school for use on the "official" Day 1.

3. *The Standard POW Write-up*

The standard POW write-up for Year 4 is the same as that for Year 3. When appropriate, the specific instructions will vary for individual POWs. In particular, be sure students notice that the write-up categories for *POW 1: The Tower of Hanoi* are somewhat different from the standard categories.

Note: Where appropriate, we will continue to have students seek to extract the mathematical essence of problems in their problem statements. That is not particularly appropriate for this POW.

> Because of the complexity of the ideas in Year 4 units, there are fewer POWs per unit in Year 4 than in previous years of the IMP curriculum. This unit contains only two POWs.

4. Introduction to *POW 1: The Tower of Hanoi*

You will probably want to take time in class to have students act out a simple version of the puzzle described in *POW 1: The Tower of Hanoi* to be sure they understand the rules. You can use the case of two discs to focus on the number of moves required. Students should see that the pile cannot be moved in fewer than three moves. You may want to let students work on the case of three discs in groups today.

Comment: Although this game is commercially available, you might let students create their own versions so that they feel more ownership of the problem.

POW 1: The Tower of Hanoi is scheduled to be discussed on Day 7.

Homework 1: The Ferris Wheel

Take a minute in class to go over the use of clock labels to represent Ferris wheel positions. (You may want to make a transparency of the blackline master of the diagram in Appendix B.)

This assignment will give students a chance to look at the Ferris wheel in a more elementary context than that of the main problem.

Simple materials combined with imagination result in a lively physical model of the unit problem.

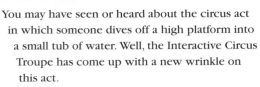

High Dive

The Circus Act

You may have seen or heard about the circus act in which someone dives off a high platform into a small tub of water. Well, the Interactive Circus Troupe has come up with a new wrinkle on this act.

They have attached the diver's platform to one of the seats on a Ferris wheel, so that it sticks out horizontally, perpendicular to the plane of the Ferris wheel. The tub of water is on a moving cart that runs along a track, parallel to the plane of the Ferris wheel, and passes under the end of the platform.

As the Ferris wheel turns, an assistant holds the diver by the ankles. The assistant must let go at exactly the right moment, so that the diver will land in the moving tub of water.

If you were the diver, would you want to trust your assistant's on-the-spot judgment? A slight error and you could get a "splat!" instead of a "splash!"

Your Task

The diver has insisted that the circus owners hire your group to advise the assistant. You need to figure out exactly when the assistant should let go. (Your analysis will be tested carefully on a dummy before it is used with a human being.)

1. Make a physical model of the problem, using materials that your teacher has provided.

Continued on next page

2. Specify any other information you need to know about the circus act to determine when the assistant should let go.

Historical note: The first Ferris wheel was created for the 1893 Chicago World's Fair and was the brainchild of George Washington Gale Ferris. This creation was much larger than the Ferris wheels of today. It stood 265 feet high and was 250 feet in diameter. It carried 36 cars, each of which could hold 60 people. A single revolution took about 20 minutes, and admission was 50 cents, ten times the cost of any other ride at the fair.

The Ferris wheel was dismantled after the fair and made brief appearances at other major events. It was sold for scrap metal in 1906.

REFERENCE

The Standard POW Write-up

The standard POW write-up for Year 4 includes the same five categories that you used in Year 3.

1. *Problem statement:* State the problem clearly in your own words. Your problem statement should be clear enough that someone unfamiliar with the problem could understand what it is that you are being asked to do.

2. *Process:* Describe what you did in attempting to solve this problem, using your notes as a reminder. Include things that didn't work out or that seemed like a waste of time. Do this part of the write-up even if you didn't solve the problem.

   If you get assistance of any kind on the problem, you should indicate what the assistance was and how it helped you.

3. *Solution:* State your solution as clearly as you can. Explain how you know that your solution is correct and complete. (If you only obtained a partial solution, give that. If you were able to generalize the problem, include your more general results.)

   Your explanation should be written in a way that will be convincing to someone else—even someone who initially disagrees with your answer.

Continued on next page

4. *Evaluation:* Discuss your personal reaction to this problem. For example, you might comment on these questions:

- Did you consider the problem educationally worthwhile? What did you learn from it?

- How would you change the problem to make it better?

- Did you enjoy working on it?

- Was it too hard or too easy?

5. *Self-assessment:* Assign yourself a grade for your work on this POW, and explain why you think you deserved that grade.

POW 1

The Tower of Hanoi

The Legend of the Golden Discs

Buddhism, one of the world's major religions, has roots in India and is practiced by over three hundred million people throughout the world. An ancient legend describes an important task once given to a group of Buddhist monks.

According to the legend, a Buddhist temple contained a pile of 64 large golden discs, one on top of another, with each successive disc slightly smaller than the one below it. This pile of discs sat upon a golden tray. Two empty golden trays lay next to the one with the pile of discs.

The monks' task was to move the pile of 64 discs from its original tray to one of the other trays. But according to the rules of the task, the monks could move only one disc at a time, taking it off the top of a pile. They could then place this disc either on an empty tray or on the top of an existing pile on one of the trays. Moreover, a disc placed on the top of an existing pile could not be larger than the disc below it.

The legend concludes with the promise that when the monks finish this task, the world will be filled with peace and harmony.

The Puzzle

As you will see, the monks could not possibly have finished the task. (How unfortunate for the state of the world!) But there is a mathematical puzzle, known as "The Tower of Hanoi," that is based on this legend. (Hanoi is the capital of Vietnam, which is located in southeast Asia. Many people in Vietnam are Buddhists.)

Continued on next page

The puzzle consists of three pegs and a set of discs of different sizes, as shown in the accompanying diagram. The discs all have holes in their centers. To begin with, the discs are all placed over the peg at the left, with the largest disc on the bottom and with the discs in decreasing size as they go up, as shown here. (This diagram uses only 5 discs instead of 64.)

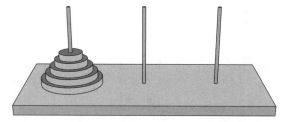

The task in this puzzle is to transfer all of the discs to the peg on the right. As in the legend, the discs must be moved according to certain rules.

• Only one disc can be moved at a time.

• The disc being moved must be the top disc on its peg.

• The disc being moved must be placed either on an empty peg or on top of a larger disc on a different peg.

Getting Started

In a sense, your POW is to answer this question.

> *If the monks had moved one disc every second, how long would it have taken them to complete the task?*

But don't start with this question. Instead, start with two or three discs and work your way up, finding the *least number* of moves that are required to transfer the pile of discs from the peg on the left to the peg on the right.

Continued on next page

As you work, consider these questions. (The notation a_n represents the number of moves required to move n discs from the peg on the left to the peg on the right.)

- If you knew how many moves were needed to move 20 discs, how could you find the number of moves needed to move 21 discs? Can you generalize this process into a formula? (That is, if you know a_n, how can you find a_{n+1}?) Can you explain why this formula holds true?

- Look for a formula that gives a_n directly in terms of n, and test your formula with specific cases. If you knew that this formula worked for $n = 20$, could you prove that it worked for $n = 21$? Can you prove the formula in general?

When you have answered these questions as best you can, go back to the question about the monks and their 64 discs.

Write-up

1. *Process*

2. *Results:* Give the results of your investigation, including

    - the number of moves required for any specific cases you studied

    - the amount of time required for the monks to move the 64 discs

    - any general formulas or procedures that you found, even if you aren't sure of them

3. *Solutions:* Explain your results, including how you know that the number of moves for each number of discs is the smallest possible. Also give any explanations you found for your generalizations.

4. *Evaluation*

5. *Self-assessment*

The Ferris Wheel

Al and Betty have gone to the amusement park to ride on a Ferris wheel. The wheel in the park has a radius of 15 feet, and its center is 20 feet above ground level.

You can describe various positions in the cycle of the Ferris wheel in terms of the face of a clock, as indicated in the accompanying diagram. For example, the highest point in the wheel's cycle is the 12 o'clock position, and the point farthest to the right is the 3 o'clock position.

For simplicity, think of Al and Betty's location as they ride as simply a point on the circumference of the wheel's circular path. That is, ignore the size of the Ferris wheel seats, Al and Betty's own heights, and so on.

1. How far off the ground are Al and Betty when they are at each of the following positions?

a. the 3 o'clock position

b. the 12 o'clock position

c. the 9 o'clock position

d. the 6 o'clock position

2. How far off the ground are Al and Betty when they are at the 2 o'clock position? (*Caution:* Their height at the 2 o'clock position is *not* a third of the way between their height at the 3 o'clock position and their height at the 12 o'clock position.)

3. Pick two other clock positions and figure out how far off the ground Al and Betty are when they reach each of those positions.

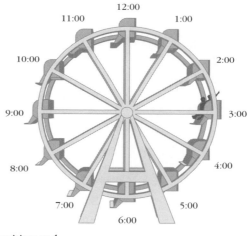

DAY 2

High-Dive Parameters

Mathematical Topics

- Reviewing basics about circles and angles
- Delineating information needed to solve the main unit problem

Outline of the Day

In Class

1. Discuss *Homework 1: The Ferris Wheel*

- Clarify the role of trigonometry in determining position on the Ferris wheel

2. Discuss *High Dive* (from Day 1)

- Have students share questions
- Give students the basic facts about the Ferris wheel setup

At Home

Homework 2: As the Ferris Wheel Turns

Special Materials Needed

- A transparency of the Ferris wheel diagram (see Appendix B)

Discuss With Your Colleagues

Where Can You Post All This Information?

This unit involves a lot of specific information and many formulas. It's helpful to students to have this information and these formulas readily available, preferably in the form of posters in the classroom. What can you do if your classroom has limited wall space or, even worse, if you don't have a classroom of your own?

1. Discussion of *Homework 1: The Ferris Wheel*

In today's discussion, students do not need to develop any general formulas concerning the relationship between position and height. They will be working with that relationship further in tonight's homework and then continue with that idea over the next week or so of the unit.

For today, it is enough that students recognize that trigonometry will play a role. You should keep in mind that from their current perspective, the trigonometric functions exist only in a right-triangle context.

Let students discuss the problems for a few minutes in their groups, and have several groups prepare presentations for examples from Questions 2 and 3.

You can probably go over the various parts of Question 1 orally. These should be straightforward if students have read the information carefully. (They may have overlooked the fact that the low point of the Ferris wheel does not have height zero. You might point out that the Ferris wheel is raised off the ground so that the seats don't scrape bottom.)

• Question 2

Then have students do their presentations from Question 2. If they have trouble with these other positions on the Ferris wheel, you will need to review right-triangle trigonometry.

For example, for the 2 o'clock position, they should see that Al and Betty's height can be found using the triangle shown in the diagram below. Thus, the height for the 2 o'clock position is given by the expression $5 + 15 + 15 \sin 30°$, which leads to a result of 27.5 feet.

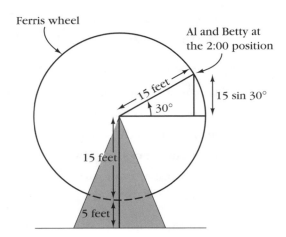

Students sometimes mistakenly assume that the heights are equally spaced from one hour to the next. For example, because the 3 o'clock position is 20 feet high and the 12 o'clock position is 35 feet high, students may think that the 2 o'clock position is a third of the way between them, at 25 feet.

If this incorrect approach is presented, you can acknowledge that it is a reasonable idea and then have the class discuss its merits. Be sure students see that it gives the wrong result. The different "o'clock" positions create equal central angles, but they do not have equally spaced heights. (If no one brings up this incorrect approach, you might want to present it yourself, to get students to articulate why it doesn't work.)

• *Question 3*

If students seemed clear about the ideas involved in their discussion of Question 2, you can omit discussion of Question 3. Otherwise, use one or two more examples to clarify the process of finding the height.

2. Discussion of *High Dive*

Have groups share their lists of questions from Question 2 of yesterday's activity. You may want to distinguish, as described here, between questions about the Ferris wheel setup and more general questions about falling objects. As the questions are proposed, you or a student can record them on chart paper.

Here are some of the questions students might ask about the setup.

- What is the radius of the Ferris wheel?

- How high is the center of the Ferris wheel above the ground?

- How fast does the Ferris wheel turn around?

- In what direction (clockwise or counterclockwise) does the Ferris wheel turn?

- When the cart starts moving, what is its position in relation to the Ferris wheel? (That is, how far is the cart from the Ferris wheel, and in what direction?)

- How fast does the cart go?

- How high is the water level in the cart above the ground?

- Where is the diver in the Ferris wheel's cycle when the cart starts moving?

(These questions may be phrased differently, or equivalent information may be requested through different questions.)

• *Facts about the Ferris wheel*

After students have posed their questions, present them with the following information about the circus Ferris wheel and the high-dive act. Give them all of this information even if they didn't specifically ask for it.

You may want to point out explicitly that the parameters for this Ferris wheel are different from those for Al and Betty's Ferris wheel from last night's homework. Tell them that the description presented here will be used throughout the unit, though occasionally they will consider other Ferris wheels.

You may want to make a poster showing the situation (or make a transparency of the diagram of the Ferris wheel in Appendix B) and mark in each detail as it is discussed. The diagram in Appendix B does not show the numerical details.

- The Ferris wheel has a radius of 50 feet.

- The center of the Ferris wheel is 65 feet off the ground.

- The Ferris wheel turns at a constant speed, making a complete turn every 40 seconds.

- The Ferris wheel turns counterclockwise.

- When the cart starts moving, it is 240 feet to the left of the center of the base of the Ferris wheel.

- The cart travels to the right at a constant speed of 15 feet per second.

- The water level in the cart is 8 feet above the ground.

- When the cart starts moving, the diver's platform is at the 3 o'clock position in its cycle.

Students should assume that when the cart starts moving, it is immediately going at 15 feet per second. (*Comment:* Students may worry that the water isn't deep enough for the diver to survive the plunge. That issue is raised again at the end of the unit.)

The picture below shows what the final diagram of the situation might look like.

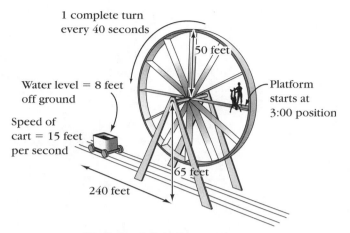

1 complete turn every 40 seconds

50 feet

Water level = 8 feet off ground

Speed of cart = 15 feet per second

Platform starts at 3:00 position

65 feet

240 feet

Portions of diagram not to scale.

Post all of the Ferris wheel and cart information prominently in the classroom, together with the diagram. Some of this information will be used right away (in tonight's homework) while other information will not be needed for a while.

Note: This is the first of many posters that you will make for this unit. Be sure to have plenty of wall space (or arrange a way to place posters in a flip-chart arrangement, which takes up less space—see today's "Discuss With Your Colleagues"). Also, encourage students to take notes on these facts and subsequent formulas, because they will be needing this information.

Students may have a variety of questions about the physics of falling objects. That is, they may wonder exactly what happens to the diver once he is released. Here are some questions that may arise.

- How fast does the diver fall?

- How long does it take the diver to reach the ground?

- Do these answers depend on the diver's weight? On his height?

Tell students that they will be learning about the mathematics and physics of falling objects later in the unit, but that they will not need to answer questions like these just yet.

• *For teachers: The diver's initial motion*

For much of the unit (through Day 16), we will be simplifying the problem by assuming that once the diver is released, he falls straight down as if he had fallen from a motionless Ferris wheel. In *Moving Cart, Turning Ferris Wheel* (begun on Day 14 and discussed on Day 16), students will solve the problem based on this assumption.

Then, after a brief interlude focusing mainly on trigonometry, students will deal with the fact that once the diver is released, his path depends not only on where he is released but also on the direction of the initial speed he gets from the motion of the turning Ferris wheel. Students begin thinking about the initial speed of the diver in *Homework 20: Initial Motion from the Ferris Wheel.*

If this complication comes up in today's discussion, you should acknowledge that the diver does not fall straight down as if from rest. Tell students that they will eventually take this into account. For now, however, they will be dealing with a simplified version of the problem in which he does fall straight down as if from rest.

Homework 2: As the Ferris Wheel Turns

In tonight's assignment, students start to look at the relationship between time elapsed, clock position, and height on the Ferris wheel.

Carol Kaneko builds a model of a Ferris wheel as an aid to better understand the unit problem.

As the Ferris Wheel Turns

In order to understand what happens when a diver is released from a moving Ferris wheel, you need precise information about the position of the diving platform as the Ferris wheel turns.

In this assignment, you will be looking only at the *height* of the platform. Later, you will consider how far the platform is to the left or right of the center of the Ferris wheel.

Continued on next page

You will need this information about the Ferris wheel.

- The radius of the Ferris wheel is 50 feet.

- The Ferris wheel turns at a constant speed, makes a complete turn every 40 seconds, and moves counterclockwise.

- The center of the Ferris wheel is 65 feet off the ground.

You should use these facts throughout the unit unless a problem specifically gives different information. *Reminder:* The circumference of a circle can be found from its radius using the formula $C = 2\pi r$.

1. At what speed is the platform moving (in feet per second) as it goes around on the Ferris wheel?

2. Through what angle (in degrees) does the Ferris wheel turn each second? (The rate at which an object turns is called **angular speed,** because it measures how fast an angle is changing. Angular speed does not depend on the radius.)

3. How many seconds does it take for the platform to go each of these distances?

   a. from the 3 o'clock to the 11 o'clock position

   b. from the 3 o'clock to the 7 o'clock position

   c. from the 3 o'clock to the 4 o'clock position

4. What is the platform's height off the ground at each of these times?

   a. 1 second after passing the 3 o'clock position

   b. 6 seconds after passing the 3 o'clock position

   c. 10 seconds after passing the 3 o'clock position

   d. 14 seconds after passing the 3 o'clock position

   e. 23 seconds after passing the 3 o'clock position

   f. 49 seconds after passing the 3 o'clock position

DAYS 3-7

The Height and the Sine

This page in the student book introduces Days 3 through 7.

Jason Weinstock presents the formula his group developed to express Al and Betty's height off the ground as the Ferris wheel turns.

You've seen that trigonometric functions can be helpful in describing where the platform is as it goes around on the Ferris wheel. But the basic right-triangle definitions of these functions only work for acute angles.

In the next portion of this unit, you'll explore how to extend the definition of the sine function to arbitrary angles and how to use this extended definition to get a general formula for the platform's height.

Students develop a formula for the platform's height when it's in the first quadrant.

A First-Quadrant Formula

Mathematical Topics

- Finding the speed of an object moving at a constant angular speed
- Finding the height, for specific times, of an object moving in a circular path, and generalizing for the first quadrant

Outline of the Day

In Class

1. Discuss *Homework 2: As the Ferris Wheel Turns*

- Post the speed of the platform as it turns
- Have students share methods for finding the platform's height at specific times

2. *At Certain Points in Time*

- Students develop a formula for the platform's height when it is in the first quadrant

3. Discuss *At Certain Points in Time*

- Post the formula for height as a function of time

At Home

Homework 3: A Clear View

1. Discussion of *Homework 2: As the Ferris Wheel Turns*

You might begin by assigning a problem to each group to prepare for presentation. Then let different spade card students report on their results.

Note: Today's activity, *At Certain Points in Time,* depends on students really understanding this homework assignment, so it's worth spending most of the class period on the homework discussion, even if that means that the activity is delayed until tomorrow.

• Questions 1 through 3

For Question 1, students will need to use the circumference formula, $C = 2\pi r$, to see that the total distance traveled by the platform in one complete turn is 100π feet. Because the platform goes 100π feet in 40 seconds, it is going 2.5π feet per second, which is approximately 7.85 feet per second (or roughly 5 miles per hour).

Post this result about the platform's speed, because it will be used later in the unit.

> **The speed of the platform as it turns is 2.5π feet per second, which is approximately equal to 7.85 feet per second.**

You may want to incorporate this information into the posted diagram of the Ferris wheel.

On Question 2, students should see that because a complete turn is 360°, the Ferris wheel must be turning $360 \div 40 = 9$ degrees each second. Review the term *angular speed* (defined in the assignment), which measures how fast an angle is changing. Bring out that angular speed is measured in units such as degrees per second.

You may want to have a student identify the angle involved, namely, the angle between the radius to the platform and some fixed radius such as the one to the 3 o'clock position. Emphasize that angular speed involves a change in direction and does not depend on the radius of the Ferris wheel.

"What word is used to describe the time interval for each complete turn of the Ferris wheel?"

Ask students what word is used to describe the time interval for each complete turn of the Ferris wheel. (You can mention the Year 1 unit *The Pit and the Pendulum* as a hint.) If necessary, remind them of the term *period*.

• Question 4

Question 4 is a lead-in to today's activity, *At Certain Points in Time,* so take extra care to ensure that students follow the presentations. You may want to post the results from Question 4, because students will use this information in later assignments.

Questions 4a and 4b are the easiest because they involve angles in the first quadrant. Use the presentations of these problems to help students work out the details of getting from the time elapsed to the angle of turn, as well as the process of using trigonometry to get the platform's height from the angle of turn. As students do various examples, they should see the need to distinguish the cases, depending on the quadrant.

For example, for $t = 6$, the situation looks like the diagram shown here. The angle is $6 \cdot 9°$ because the Ferris wheel turns 9 degrees per second. Students will probably use the equation $\sin 54° = \frac{y}{50}$ to get $y = 50 \sin 54°$. Therefore, the total height off the ground is $h = 65 + 50 \sin 54°$.

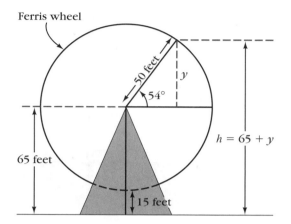

Students may recognize that Question 4c is the same as Question 1b of *Homework 1: The Ferris Wheel* except for the conversion from time to degrees.

For Question 4d ($t = 14$), the platform will have moved through an angle of $126°$. Students may recognize that this leads to the same height for the platform as for Question 4b ($t = 6$). If they do not see this, they will probably use either the supplementary angle of $54°$ (getting $65 + 50 \sin 54°$ for the height of the platform) or the related $36°$ angle (getting $65 + 50 \cos 36°$). These two approaches are illustrated in the two diagrams on the next page. You don't need to discuss both methods, but, as always, encourage alternate approaches to problems.

Note: If both approaches come up, you can bring out that $\sin 54° = \cos 36°$ and discuss why. (The general principle that $\sin \theta = \cos (90° - \theta)$ was first discussed in the IMP curriculum in *Homework 24: Your Opposite Is My Adjacent* in the Year 1 unit *Shadows,* but it may need review.)

Also, students may simply follow the pattern of Questions 4a and 4b, and use the expression $65 + 50 \sin 126°$. They may discover that this expression yields a reasonable answer on their calculators, even though $\sin 126°$ has not yet been defined. If this comes up, remind them that so far, they only know the meaning of the trigonometric functions for acute angles. Then tell them that extending the definitions of these functions is an important element in this unit.

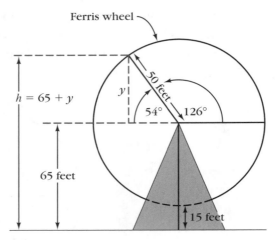

Here, we have $\frac{y}{50} = \sin 54°$, so $y = 50 \sin 54°$ and $h = 65 + 50 \sin 54°$.

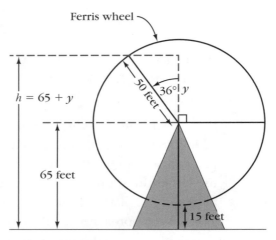

Here, we have $\frac{y}{50} = \cos 36°$, so $y = 50 \cos 36°$ and $h = 65 + 50 \cos 36°$.

For Question 4f, students will probably realize that when the Ferris wheel has turned for 49 seconds, the platform is at the same position as after 9 seconds. Use the word **periodic** to describe the motion of the platform.

2. At Certain Points in Time

In this activity, students will develop a general formula *for the first quadrant* for the homework situation.

3. Discussion of *At Certain Points in Time*

Let a heart card student present and explain his or her group's formula. This should follow fairly easily from the homework discussion, but be sure to get a clear explanation, because this formula will play a major role in extending the sine function beyond acute angles.

Post and label the formula, which will probably look like this:

$$h = 65 + 50 \sin 9t$$

You can tell students that they will consider other quadrants tomorrow.

Homework 3: A Clear View

This assignment uses Al and Betty's Ferris wheel and poses a question that isn't explicitly about only height or time.

Kit Taggert and Amanda Matthews review whether their formula accurately expresses Al and Betty's height off the ground as the Ferris wheel turns.

At Certain Points in Time

In *Homework 2: As the Ferris Wheel Turns,* you found the height of the platform after it had turned for specific amounts of time. You probably saw that this was easiest to do if the platform was in the first quadrant.

Your task in this activity is to generalize that work for the case of the first quadrant. (*Reminder:* The basic facts about the Ferris wheel are the same as in *Homework 2: As the Ferris Wheel Turns.* In particular, the period is 40 seconds, so the platform remains in the first quadrant for the first 10 seconds.)

1. Suppose the Ferris wheel has been turning for t seconds, with $0 < t < 10$. Represent the platform's height off the ground as $h,$ and find a formula for h in terms of t.

2. Verify your "first-quadrant formula" using your results from Questions 4a ($t = 1$) and 4b ($t = 6$) of *Homework 2: As the Ferris Wheel Turns.*

A Clear View

As you may remember, the Ferris wheel at the amusement park where Al and Betty like to ride has a radius of 15 feet, and its center is 20 feet above ground level. (This is not the same Ferris wheel as the one at the circus.)

The Ferris wheel turns with a constant angular speed and takes 24 seconds for a complete turn.

There is a 13-foot fence around the amusement park, but once you get above the fence, there is a wonderful view.

1. What percentage of the time are Al and Betty above the height of the fence? (You may want to find out how long they are above the height of the fence during each complete turn of the Ferris wheel.)

2. How would the answer to Question 1 change if the period were different from 24 seconds?

DAY 4

Students extend the definition of the sine function to all angles.

Beyond the First Quadrant

Mathematical Topics

- Extending the sine function to be defined for all angles

Outline of the Day

In Class

1. Discuss *Homework 3: A Clear View*

2. Discuss the need to extend the trigonometric functions

- Bring out that the sine function has previously been defined only for first-quadrant angles
- Have students do focused free-writing on the idea of extending the definition of exponentiation
- Use a specific Ferris wheel situation to illustrate the extended definition

3. Formally define the extended sine function

- Confirm that this definition is consistent with the specific case already discussed and that calculators agree with the new definition for the specific case

4. Optional: Have students read *Extending the Sine*

5. *Testing the Definition*

- Students test the extended definition of the sine function using the platform height function
- The activity will be discussed on Day 5

At Home

Homework 4: Graphing the Ferris Wheel

1. Discussion of *Homework 3: A Clear View*

Ask for a volunteer to present the problem. Be sure that the presenter includes a diagram as part of the explanation. For instance, the diagram here shows the situation when Al and Betty are exactly 13 feet off the ground (and in the fourth quadrant). At this position, they are 7 feet below the center of the Ferris wheel, so the angle θ satisfies the equation $\cos \theta = \frac{7}{15}$, which gives $\theta \approx 62.2°$. (You can use this discussion, if needed, as an opportunity to remind students of the inverse trigonometric functions and notation such as $\cos^{-1} \frac{7}{15}$.)

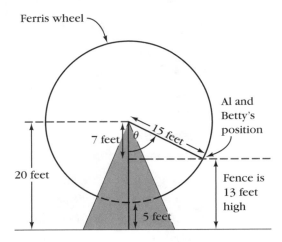

The presenter might use a diagram like the next one to bring out that Al and Betty are below the fence while the Ferris wheel travels through an angle of *twice* 62.2°, using the fact that there are two places in the Ferris wheel cycle where Al and Betty are 13 feet high.

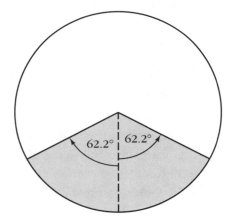

Some students may work directly with the angles. For instance, they might see that the fraction of the time during which Al and Betty are *below* the fence is equal to the ratio $\frac{2 \cdot 62.2}{360}$, or approximately 35%. Because Al and Betty are below 13 feet about 35% of the time, they can see over the fence about 65% of the time.

Some students may feel more comfortable working with the time involved. For instance, they might see that each second elapsed represents 15 degrees of turn, so 62.2 degrees represents $\frac{62.2}{15}$ seconds, or about 4.15 seconds. This means that Al and Betty are below the fence for about 8.3 seconds and above it for about 15.7 seconds. Thus, the fraction of the time that they are above the fence is about $\frac{15.7}{24}$, or approximately 65%.

You can use Question 2 to bring out that although the period of the Ferris wheel affects the *amount* of time in each cycle that Al and Betty are above the fence, it does not affect the *percentage* of time they are above the fence.

Comment: There are several ways to do Question 2. As usual, be sure to encourage alternate explanations. If some students used the period of the Ferris wheel in working on Question 1, you might have them explain how their use of the value 24 "canceled out" in their arithmetic. For instance, they may have multiplied by 24 in one part of the computation and divided by 24 in another part.

2. Extending the Sine Function

Review yesterday's work developing a first-quadrant formula for the platform's height above the ground, namely

$$h = 65 + 50 \sin 9t$$

(You may need to clarify that the Ferris wheel in the unit problem has different parameters from that in the homework.)

"Why can't you simply apply this formula for all values of t?"

Point out that it would be nice to have a formula that works no matter where the platform is, and ask why students can't simply apply this formula for all values of t. Bring out that the definition of the sine function is based on right triangles, so the expression $\sin 9t$ is meaningful only if $9t$ is strictly between 0° and 90°. That is, the formula makes sense (so far) only if the platform is still in the first quadrant, which means $0 < t < 10$. *(Note:* Even if students have already discovered that the formula seems to work for other angles, point out that they don't have a general definition yet for the sine function.)

• *Previous experiences with extending functions*

"When have you extended a function or operation before?"

Tell students that their next task in the unit is to consider how to extend the sine function to be defined for all angles. Ask them what other experiences they can recall in which they extended a function or operation to include other numbers. Remind them, if necessary, that when they first learned about exponentiation, the operation was defined in terms of repeated multiplication and that the definition made sense only if the exponent was a positive integer.

Then ask students to review how they extended exponentiation in the Year 2 unit *All About Alice.* Try to bring out these key ideas.

- The new definition extending the domain of the operation had to be consistent with the old definition.

- The situation of Alice and the cake and beverage was a useful model for thinking about exponents.

- The new definition was created so that certain patterns and algebraic rules that held true for positive integer exponents continued to hold true when the domain was extended.

Tell students that the model of the Ferris wheel for circular motion will play a role for trigonometric functions similar to that played by the Alice situation for exponentiation. You might also tell students that they will probably not find any numerical patterns as simple as those they used for exponentiation.

You might mention as well that circular motion similar to that of the Ferris wheel occurs in many contexts, so generalizing the formula for the height of the platform will be useful in other situations as well.

• A specific case

Tell students that a key criterion of a new, extended definition of the sine function will be whether it makes the platform height function work for all values of t.

*"What value should
you assign as the
definition of sin 126°
so that the height
formula gives the right
answer for t = 14?"*

Point out that when $t = 14$, the height formula gives the expression $65 + 50 \sin (9 \cdot 14)$, which means that the angle of turn for the platform is 126° and the platform would no longer be in the first quadrant. Ask what value students would need to assign as the definition of sin 126° so that the formula would give the right value for the platform's height at $t = 14$.

*As a hint: "What value
of t gives the same
height but leads to a
first-quadrant angle?"*

As a hint, you can ask what value of t gives the same height as $t = 14$ but keeps the platform in the first quadrant. Bring out that at $t = 14$, the platform has the same height as at $t = 6$, so the expression $65 + 50 \sin (9 \cdot 14)$ must come out the same as the expression $65 + 50 \sin (9 \cdot 6)$. To make this more vivid, you might write this as the equation

$$65 + 50 \sin 126° = 65 + 50 \sin 54°$$

Thus, for the height formula to work for $t = 14$, we must define the sine function so that sin 126° has the same value as sin 54°. (This relationship is more important here than the numerical value for sin 126°.)

3. Defining the Sine Function for All Angles

Tell students that in order to create a general, extended definition of the sine function, it's helpful to replace the Ferris wheel model with the more abstract setting of the coordinate plane. You can begin with a diagram like the one shown here, suggesting that they imagine that the circle of the Ferris wheel has been placed in the coordinate plane, with its center at the origin.

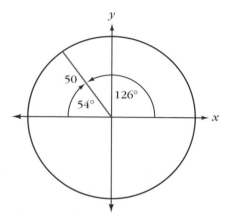

Then ask the class to consider the next diagram, which corresponds to a situation in which the Ferris wheel has turned only through a first-quadrant angle. Some point *A*, with coordinates (x, y), has been marked on the ray defining the angle. (The point *A* is assumed to be different from the origin.)

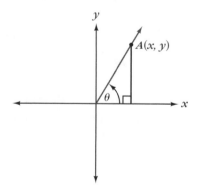

"How can sin θ be defined in terms of the coordinates x and y?"

Ask how $\sin \theta$ could be defined in terms of the coordinates x and y. Students should see that in the right triangle, $\sin \theta$ is equal to the ratio $\frac{y}{\sqrt{x^2 + y^2}}$.

Introduce the use of r as a shorthand for the expression $\sqrt{x^2 + y^2}$, and bring out that this corresponds to the radius of the Ferris wheel. (*Note:* Polar coordinates will be introduced on Day 17.) Use the variable r to rewrite the expression for $\sin \theta$ more simply as the ratio $\frac{y}{r}$.

Point out that the ratio $\frac{y}{r}$ makes sense for any angle. (The issue of negative angles will be discussed explicitly tomorrow.) Tell students that this simple ratio is used for the extended definition of the sine function. Post the formal definition together with an appropriate diagram:

For any angle θ, we define sin θ by first drawing the ray that makes a counterclockwise angle θ with the positive x-axis and choosing a point A on this ray (other than the origin) with coordinates (x, y).

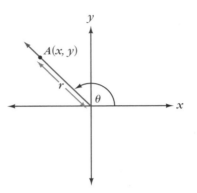

Using the shorthand $r = \sqrt{x^2 + y^2}$, we then define the sine function by the equation

$$\sin \theta = \frac{y}{r}$$

Note: The issue of why this method makes sin θ well defined is discussed on Day 5 (see the section "Why Is the Sine Function Well Defined?"). Unless students ask why the ratio $\frac{y}{r}$ comes out the same for all points on the ray, it is probably best not to complicate things by discussing this now.

This extension of the right-triangle sine function to the more general definition of sine for arbitrary angles is a key idea, and perhaps even the central mathematical idea of this unit, so be sure to give it appropriate fanfare.

Bring out that we have taken familiar ideas — right-triangle trigonometry and the coordinate system — and combined them in the context of a concrete situation to create a more general definition of the sine function. This new definition is consistent with the old definition of sine for acute angles. It allows us to replace a complex, quadrant-by-quadrant analysis of the platform height with a single, uniform expression.

You can tell students that this new definition fits other situations and mathematical rules that they have not yet encountered.

• *Back to the specific case*

Have students work in groups to see how this definition applies to the angle of 126° that they worked with earlier. You may need to suggest that they pick a specific value for *r* (perhaps *r* = 50, as in the Ferris wheel) and then find the corresponding value for *y,* as in this diagram.

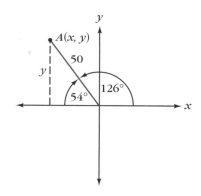

Students should see, perhaps using the right triangle in the second quadrant, that *y* is equal to 50 sin 54°, so the ratio $\frac{y}{r}$ is equal to sin 54°. In other words, this coordinate definition will make sin 126° the same as sin 54°, as before.

Have students verify that their calculators do, in fact, give the same value for sin 126° as for sin 54°.

4. For Reference: *Extending the Sine*

The material here summarizes the ideas just discussed for extending the sine function. You may want to have students look over this briefly, or you might have them begin work directly on the next activity, *Testing the Definition.*

5. *Testing the Definition*

Groups can begin work right away on this activity, which will be discussed on Day 5. (You need not have all groups complete Question 2 before beginning discussion.)

Question 1 uses the special case in which the right triangle has a 45° angle. On Question 2, students will need to use right-triangle trigonometry to find the coordinates of point *A* in the fourth quadrant.

Homework 4: Graphing the Ferris Wheel

You might suggest to students that they use their data from *Homework 2: As the Ferris Wheel Turns* in Question 1 of tonight's homework. Question 2 represents an intuitive look at how changing the parameters affects the graph. Students will look at this in more detail in *Homework 5: Ferris Wheel Graph Variations.*

Extending the Sine

If the Ferris wheel turns counterclockwise at a constant angular speed of 9 degrees per second and the platform passes the 3 o'clock position at $t = 0$, then the platform will remain in the first quadrant through $t = 10$.

During this time interval, the platform's height above the ground is given by the formula

$$h = 65 + 50 \sin 9t$$

But the right-triangle definition of the sine function makes sense only for acute angles. To make this formula work for all values of t, we need to extend the definition of the sine function to include all angles.

The Coordinate Setting

Although the context of the Ferris wheel could be used to develop this extended definition, the standard approach uses a more abstract setting, which makes it easier to apply the definition to other situations.

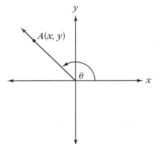

Specifically, the angle is placed within a coordinate system, with its vertex at the origin, and is measured counterclockwise from the positive direction of the x-axis. The goal is to express $\sin \theta$ in terms of the x- and y-coordinates of a point on the ray defining the angle, such as the point A shown in the first diagram. (A is assumed to be different from the origin.)

When θ is an acute angle, we get a diagram like the second one. It's helpful to introduce the letter r to represent the distance from A to the origin, which is also the length of the hypotenuse of the right triangle.

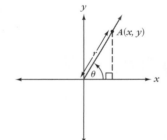

Continued on next page

Interactive Mathematics Program

Based on the right-triangle definition of the sine function, we get

$$\sin \theta = \frac{y}{r}$$

Mathematicians use this equation to extend the definition of the sine function to arbitrary angles. That is, they define $\sin \theta$ as the ratio $\frac{y}{r}$ for *any* angle θ. (*Comment:* This automatically means that the new definition agrees with the old one for acute angles.)

The Ferris Wheel Analogy

You can think of A as a point on the circular path of the Ferris wheel, as shown in this diagram. In this context, r corresponds to the radius of the Ferris wheel, and y corresponds to the platform's height *relative to the center of the Ferris wheel*.

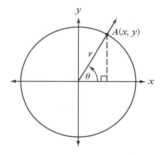

Testing the Definition

You've seen that the sine function can be extended to all angles using the xy-coordinate system. The big question is this.

If you use this coordinate definition of the sine function, does the platform height formula work for all angles?

Your task in this activity is to investigate that question.

1. If the platform has been turning for 25 seconds, then it has moved through an angle of 225° and is now in the third quadrant of its cycle.

a. Use a diagram like the one shown here to find the value of sin 225° based on the coordinate definition of the sine function. (*Suggestion:* Choose a specific value for r. Then find the value of y using the right triangle in the third quadrant.

b. Substitute your answer from Question 1a into the expression 65 + 50 sin 225°.

c. Explain why your answer in Question 1b is a reasonable answer for the position of the platform after 25 seconds.

d. Verify that your calculator gives the same value for sin 225° that you found in Question 1a.

2. Go through a sequence of steps like those in Question 1, but use the value $t = 32$, which places the platform in the fourth quadrant. (You will first need to find the actual height of the platform for $t = 32$.)

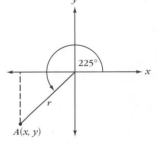

Graphing the Ferris Wheel

1. Plot individual points to create a graph showing the platform's height, *h,* as a function of the time elapsed, *t.* Explain how you get the value for *h* for each point you plot. Your graph should show the first 80 seconds of the Ferris wheel's movement.

Reminder: Use the same basic information about the Ferris wheel as in *Homework 2: As the Ferris Wheel Turns.*

2. Describe in words how this graph would change if you made each of the changes described in Questions 2a through 2c. Treat each question as a separate problem, changing only the item mentioned in that problem and keeping the rest of the information as in Question 1.

a. How would the graph change if the radius of the Ferris wheel was smaller?

b. How would the graph change if the Ferris wheel was turning faster (that is, if the period was shorter)?

c. How would the graph change if you measured height with respect to the center of the Ferris wheel instead of with respect to the ground? (For example, if the platform was 40 feet above the ground, you would treat this as a height of −25, because 40 feet above the ground is 25 feet below the center of the Ferris wheel.)

DAY 5

Students work with the extended definition of the sine function.

Extending the Sine Function

Mathematical Topics

- Introducing the term **reference angle**
- Establishing a general formula for the height of an object moving on a Ferris wheel as a function of time
- Using similarity to see that the sine function is well defined
- Graphing the Ferris wheel height function

Outline of the Day

In Class

1. Discuss *Testing the Definition* (from Day 4)
  - Examine the specific cases as needed
  - Summarize the connection between the sine function and the Ferris wheel height function
  - Discuss the case of negative angles
  - Discuss the sign of the sine function

2. Have students use similarity to demonstrate that the sine function is well defined

3. Introduce the unit circle

4. Discuss *Homework 4: Graphing the Ferris Wheel*
  - Bring out the periodicity of the graph

At Home

Homework 5: Ferris Wheel Graph Variations

Special Materials Needed

- Transparencies of the blank and completed graphs for discussion of *Homework 4: Graphing the Ferris Wheel* (see Appendix B)

Note: We recommend that you discuss yesterday's activity, *Testing the Definition*, and other ideas related to the extension of the definition of the sine function before discussing the homework.

1. Discussion of *Testing the Definition*

Let diamond card students from different groups present different parts of Question 1. For Question 1a, students might pick 50 for r and see that the legs of the right triangle are both of length $25\sqrt{2}$. They might get this by using the Pythagorean theorem, by using trigonometric functions of a 45° angle, or simply by remembering the ratios for special case of an isosceles right triangle.

Once students have the lengths of the sides, they need to see that y is negative, because point A is in the third quadrant. (*Comment:* They don't actually need to find the value of x to get the value of sin 225°, but you might have them do this anyway.)

Finally, students need to find the ratio $\frac{y}{r}$ and see that this ratio is approximately 0.707. (If they leave the ratio in terms of $\sqrt{2}$, that's fine, too.)

- *Questions 1b through 1d*

When students substitute their value for sin 225° into the expression $65 + 50 \sin 225°$, they should get a value of approximately 29.6 feet. Their explanation that this is reasonable should certainly include the fact that the result is less than 65 feet (because the platform is below the center of the Ferris wheel) but more than 15 feet (because the lowest point in the cycle is 15 feet off the ground).

You might ask them to find the position of the platform after 5 seconds (which corresponds to an angle of 45°). Then bring out that the result there (about 100.4 feet) is about 35.4 feet above the center of the Ferris wheel, just as 29.6 feet is 35.4 feet below the center.

Be sure to have students see that their calculators give the same value for sin 225° as the value they just found.

Comment: Students will be discussing shortly why it doesn't matter which point on the ray is chosen—see the next section, "Why Is the Sine Function Well Defined?"

- *Question 2*

"How can you express sin 288° in terms of the sine of a first-quadrant angle?"

Question 2 is similar to Question 1, except that students will likely have more difficulty finding the value of y here than in Question 1. Ask how students might express sin 288° in terms of the sine of a first-quadrant angle. They should see that except for sign, sin 288° is like sin 72°, so sin 288° $= -\sin 72°$. As with Question 1, have them see that their calculators give this result also.

- *The reference angle*

Bring out that for any angle, there is a first-quadrant angle that has the same sine, except perhaps for the sign. (Be careful to distinguish between the words "sign" and "sine.")

Tell students that this first-quadrant angle is called the **reference angle.** They should see that the reference angle for 225° (from Question 1) is 45° and that the reference angle for 288° is 72°.

- *The general platform height function*

Assure students that with the coordinate definition of the sine function, their platform height function works for all values of *t*. Whenever the platform is above the center of the Ferris wheel, *y* is positive; whenever the platform is below the center of the Ferris wheel, *y* is negative. Thus, the platform's overall height is 65 + *y*, regardless of the sign of *y*.

By the general definition, $\sin \theta = \frac{y}{r}$, so $y = r \sin \theta$. In the case of the Ferris wheel in the unit problem, *y* becomes 50 sin θ. In other words, the height of the Ferris wheel is always 65 + 50 sin θ. That is the power of using the coordinate plane to understand the Ferris wheel.

Post this general result, which you might state as follows.

Suppose the Ferris wheel and platform satisfy these conditions:

- **The center of the Ferris wheel is 65 feet off the ground.**

- **The radius of the Ferris wheel is 50 feet.**

- **The Ferris wheel turns counterclockwise at a constant rate with a period of 40 seconds.**

- **The platform is at the 3 o'clock position when *t* = 0.**

Then the height of the platform off the ground after *t* seconds is given by the expression

$$65 + 50 \sin 9t$$

Be sure students see how the specific details of the circus act fit into this formula. In particular, you may want to go over the fact that the coefficient 9 (in the expression 9*t*) comes from dividing 360° by the period of 40 seconds. You might review the term *angular speed* by bringing out that the angular speed of the Ferris wheel is 9 degrees per second.

• Negative angles

"How might the extended definition of the sine function make sense when θ is negative?"

Ask how the extended definition of the sine function might make sense if θ is negative. You may want to suggest that students consider what a negative value for t would mean in the context of the Ferris wheel problem.

Students may be able to guess on their own, but if needed, tell them that we interpret a negative angle by going clockwise instead of counterclockwise and that a negative value of t refers to a time before the platform reaches the 3 o'clock position.

Illustrate with a specific case, such as asking students to find $\sin(-53°)$. They should see that the reference angle is 53° and that $\sin(-53°) = -\sin 53°$.

> Students will develop the general formula $\sin(-\theta) = -\sin\theta$ in the next unit, *As the Cube Turns.*

2. Why Is the Sine Function Well Defined?

> The definition of $\sin\theta$ involves the use of a point on the ray making an angle of θ measured counterclockwise from the x-axis. The question may have come up earlier as to why we can use *any* point on the ray. If it has not yet been discussed, bring it up now.

Review the fact that $\sin\theta$ is defined by the ratio $\frac{y}{r}$ for some point (other than the origin) on the ray defining θ. You might use a diagram like the one shown here.

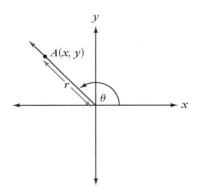

"Why is the ratio $\frac{y}{r}$ the same for all points on the ray?"

Then ask why the ratio is the same for all points on the ray. If necessary, revise the diagram to show two points on the ray, perhaps labeled (x_1, y_1) and (x_2, y_2), with r-values r_1 and r_2, as in the next diagram.

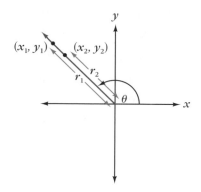

"Why is the ratio $\frac{y_1}{r_1}$ the same number as $\frac{y_2}{r_2}$?"

If needed, restate the question in terms of the new diagram, asking why the ratio $\frac{y_1}{r_1}$ should be the same number as $\frac{y_2}{r_2}$.

The goal here is to bring out that similarity plays a key role in this general definition, just as it did when the trigonometric functions were defined for acute angles using right triangles. You might point out that no matter what quadrant θ is in, the two ratios will have the same sign because the y-coordinates of the two points have the same sign. Students can then use similar right triangles to see that the ratios of the lengths of the sides are the same by similarity.

3. The Unit Circle

"What value for r would be simplest to use?"

Tell students that because the value of the sine does not change for different values of r, they can choose whatever value is most convenient. Ask what value might make things simplest. Bring out that using 1 for r means simply that $\sin \theta = y$.

"How can you describe the set of points with $r = 1$?"

Ask students to describe the set of points with $r = 1$. They should see that these points form a circle of radius 1, with the center at the origin. Tell them that this set of points is called the **unit circle.**

4. Discussion of *Homework 4:* *Graphing the Ferris Wheel*

"How should the axes and scales be set up?"

Ask for a volunteer to discuss how to set up the axes and scales for the graph in Question 1. The vertical scale should reflect the fact that the height goes from a minimum of 15 feet to a maximum of 115 feet. Students are specifically instructed to include a horizontal scale from $t = 0$ to $t = 80$.

We suggest that you set up a transparency with these scales so that students can plot their points on a shared coordinate system. (A blackline master of a blank coordinate system with these scales is included in Appendix B.) You can then have club card students from various groups each give and explain the coordinates for one point on the graph. You can label various points with the expressions used to find the values for *h*.

Some students may be finding the heights in terms of right triangles, just as they did at the beginning of the unit, and perhaps even using the cosine of some angle rather than the sine. If so, you may get a diagram with points labeled as shown here. (Appendix B contains a larger version of this graph without individual points labeled.)

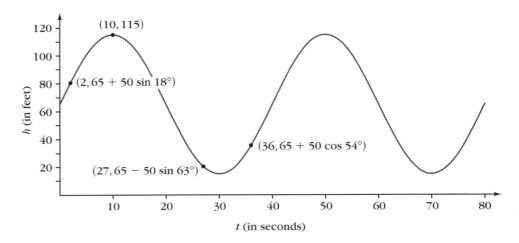

Ask what single formula will describe this graph. If needed, review the discussion from earlier today, so students see that this is the graph of the function $h = 65 + 50 \sin 9t$.

Have students graph this function on their calculators (adjusting the variables as needed for calculator entry and choosing an appropriate viewing window). They should see that the graph matches the one they created by hand.

Note: Some students are likely to have more of a "sawtooth" graph. Be sure they see that the graph is a smooth curve with no "corners." You might use the Ferris wheel itself as a way of explaining why this is so.

• *Symmetry and periodicity in the graph*

"Did you use any shortcuts or see any patterns?"

Ask students if they used any shortcuts or saw any patterns in their results. It would be nice if they mention things like symmetry or periodicity here.

"What does it mean that f is periodic, with period 40 seconds?"

This is a good time to review the idea of periodicity. Ask students to describe in words what it means that f is periodic, with period 40 seconds. They might say something like, "The height is the same every 40 seconds." You can have a student use the graph to illustrate what this means, by showing that points whose t-coordinates differ by 40 have the same h-coordinate.

Bring out that when we say that the period for the height function is 40 seconds, this means not only that the height is the same every 40 seconds but also that there is no smaller time interval for which the height always repeats.

• Question 2

The discussion of Question 2 should be limited to a qualitative description of the changes in the graphs. In tonight's assignment, students will look at the graphs for specific variations of each type.

In discussing Question 2a, students should be able to explain that if the radius were smaller, the new graph would not go as "high" or as "low" as the original. They might describe the new graph as "squished vertically toward the line $y = 65$." Bring out that the "midline" of the graph remains the same. That is, the graph is still as much above the line $y = 65$ as it is below this line.

Tell students that the distance from the midline to the high or low point of the graph is called the **amplitude** of the graph. In other words, the amplitude for such a Ferris wheel height graph is the same as the radius of the Ferris wheel.

On Question 2b, students should see that if the Ferris wheel turns faster, then the platform will go up and down more times during the 80-second interval shown. In other words, the height function will have a smaller period. They might describe the graph as "squished horizontally like an accordion."

"How is the amplitude affected in Question 2c?"

Finally, on Question 2c, students should see that the graph is simply moved down so that it is above the axis half the time and below the axis half the time. Ask what has happened to the amplitude. Students should see that the amplitude has not changed.

Homework 5: Ferris Wheel Graph Variations

In this assignment, students examine how changing the specifications of the Ferris wheel affects the graph of the platform's height.

Ferris Wheel Graph Variations

In Question 1 of *Homework 4: Graphing the Ferris Wheel*, you made a graph showing how the height of the platform depends on the amount of time that has elapsed since the Ferris wheel began moving. That graph was based on the "standard" Ferris wheel, which has a radius of 50 feet and a period of 40 seconds and whose center is 65 feet off the ground. The accompanying diagram shows two periods of that graph.

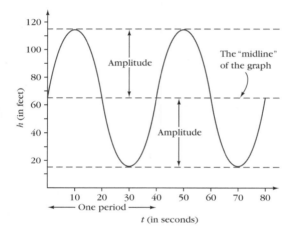

The dashed line at $h = 65$ shows the "midline" of the graph. The graph is as much above this line as it is below. The other dashed lines, at $h = 115$ and $h = 15$, show the maximum and minimum of the graph. The distance from the midline to the maximum or minimum is called the **amplitude** of the graph, so the amplitude here is 50.

In Question 2 of *Homework 4: Graphing the Ferris Wheel*, you described in words how the graph would

Continued on next page

change if you made certain changes in the Ferris wheel itself. In this assignment, you will look at those changes in more detail.

Treat each question as a separate problem, changing only the item mentioned in that problem and keeping the rest of the information as in the standard Ferris wheel. Use the same scale for all of your graphs so that you can make easy visual comparisons.

1. **a.** Pick a specific new value, less than 50 feet, for the radius, and draw the graph.

   b. Give an equation for your new graph, expressing h (the height of the platform, in feet) in terms of t (the time elapsed, in seconds).

   c. Pick a specific value for t and verify that your equation from Question 1b gives the value you used in your graph for that value of t.

2. **a.** Pick a specific new value, less than 40 seconds, for the period, and draw the graph.

   b. Give an equation for your new graph, expressing h in terms of t.

   c. Pick a specific value for t and verify that your equation from Question 2b gives the value you used in your graph for that value of t.

3. Suppose the Ferris wheel were set up inside a large hole so that its center was exactly level with the ground.

   a. Draw the graph based on this change.

   b. Give an equation for your new graph, expressing h in terms of t.

   c. Pick a specific value for t and verify that your equation from Question 3b gives the value you used in your graph for that value of t.

DAY 6 Sine Graphs

Mathematical Topics

- Examining how the graph of the Ferris wheel height function varies as parameters are changed
- Examining the graph of the "plain" sine function

Students study the graphs of the Ferris wheel height function and the "plain" sine function.

Special Materials Needed

- A transparency of the graph of the function $h = \sin t$ (see Appendix B)

Outline of the Day

In Class

1. Select presenters for tomorrow's discussion of *POW 1: The Tower of Hanoi*

2. Discuss *Homework 5: Ferris Wheel Graph Variations*

- Discuss how the graph of the Ferris wheel height function varies as parameters are changed

3. *The "Plain" Sine Graph*

- Students examine in detail the graph of the function $h = \sin t$

4. Discuss *The "Plain" Sine Graph*

- Go over the graph's amplitude, period, intercepts, and maxima and minima
- Relate the "plain" sine function to the Ferris wheel problem

At Home

Homework 6: Sand Castles

---

1. POW Presentation Preparation

Presentations of *POW 1: The Tower of Hanoi* are scheduled for tomorrow. Choose three students to make POW presentations, and give them overhead transparencies and pens to take home to use for preparing presentations.

2. Discussion of *Homework 5: Ferris Wheel Graph Variations*

You can have volunteers present specific graphs for each of the questions. Keep in mind that other students may have chosen different values for the radius or period, so students will probably have slightly different graphs. You can use the specific examples to review the general ideas that were included in the discussion of *Homework 4: Graphing the Ferris Wheel*.

• Question 1

"How does the amplitude of this graph compare to that of the original graph?"

If the presenter of Question 1 does not use the term *amplitude,* ask the class specifically how the amplitude of this graph compares to that of the graph shown in the assignment. Students should see that the amplitude of the graph for Question 1 is less than that of the "original" platform height graph (from *Homework 4: Graphing the Ferris Wheel*).

"What does the amplitude depend on?"

More generally, ask what the amplitude for a Ferris wheel height graph depends on. Bring out that the amplitude of the graph is equal to the radius of the Ferris wheel.

Connect the changes in the Ferris wheel and the graph to the change in the equation. For instance, if a student changed the radius to 30 feet, the equation would be $h = 65 + 30 \sin 9t$.

"Has anything happened to the period?"

Ask how the period of this new graph compares to that of the original graph. Students should see that the period is not affected by the change in radius.

• Question 2

"Has anything happened to the amplitude?"

For Question 2, ask how the amplitude of this new graph compares to that of the original graph. Students should see that the amplitude is not affected by the change in period.

As with Question 1, connect the change in the Ferris wheel and the graph to the change in the equation. For instance, if a student changed the period to 20 seconds, the coefficient of t would change from 9 to 18 (found by dividing $360°$ by 20), and the equation would be $h = 65 + 50 \sin 18t$.

• Question 3

For Question 3, students should all have the same graph and equation. They should see that the entire graph is merely moved down so that the x-axis is its midline. The equation is simply $h = 50 \sin 9t$.

"Has anything happened to the period or amplitude?"

Ask how the period and amplitude of this graph compare to those of the original graph. Students should see that they have not changed.

In the supplemental problem *A Shifted Ferris Wheel,* students investigate the effect of changing the initial location of the platform from the 3 o'clock position to something else.

3. The "Plain" Sine Graph

In the last two homework assignments, students worked with graphs of "Ferris wheel situations" involving the sine function. In the next activity, *The "Plain" Sine Graph,* students will examine the graph of the sine function itself.

4. Discussion of *The "Plain" Sine Graph*

The discussion of this activity can be brief if students were comfortable answering the specific questions in their groups.

You should post a copy of the graph itself. (The diagram shown here is included in Appendix B, in case you want to use a transparency of the graph in the discussion.)

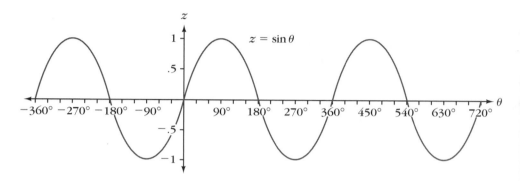

- ## Some further details

  Here are some specific connections and ideas to bring out, if they do not seem clear from students' group work.

  - Have students explain in terms of the coordinate definition why the sine function has a maximum of 1 and a minimum of −1.

  - Bring out the connection between the fact that the amplitude of the function is 1 and the fact that the sine function has a maximum of 1 and a minimum of −1.

  - Have students write an equation expressing the fact that the period of the function is 360° (for instance, they might write $\sin(\theta + 360°) = \sin\theta$) and explain this fact in terms of the coordinate definition.

  - Have students relate the θ-intercepts to the coordinate definition of the sine function. They should see that the sine function is zero when the "defining point" (that is, the point on the appropriate ray) has a y-coordinate of zero, which means that the point is on the x-axis.

  - Bring out that the angles that are involved in the intercepts, maximum points, and minimum points are precisely the angles for which the defining point is on one of the axes, so that y is either zero or is equal to r in absolute value.

- ## *The sign of the sine*

"For what angles is the sine function positive? When is it negative?"

Ask students to generalize about the sign of the sine function. By looking at the sign of the *y*-coordinate, students should be able to determine that the sine function comes out positive if the angle is in the first or second quadrant and negative if the angle is in the third or fourth quadrant. (Bring out that an angle such as $-53°$ is a fourth-quadrant angle.)

You may find it helpful to use a diagram like the one shown here.

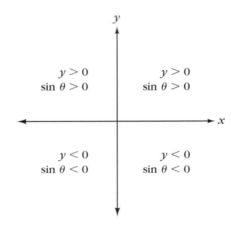

- ## *The sine function and the Ferris wheel*

"What specifications for the Ferris wheel would give this graph?"

Finally, bring out that this graph has the same basic shape as the "height functions" that students examined in *Homework 5: Ferris Wheel Graph Variations*. Ask what specifications for the Ferris wheel would give this graph. They should see that this graph shows the height of the platform for a Ferris wheel with a radius of 1 unit and an angular speed of 1 degree per second and in which the center of the wheel is at ground level.

Homework 6: Sand Castles

This assignment illustrates the use of the sine function to describe periodic motion in a context that is quite different from the Ferris wheel and that does not involve angles.

The "Plain" Sine Graph

The height of the Ferris wheel platform is given by a formula that involves the sine function. In previous assignments, you've graphed this height function and examined how the graph changes as various details of the Ferris wheel itself are changed.

In this activity, you'll look at the graph of the "plain" sine function.

1. Draw the graph of the function defined by the equation $z = \sin \theta$ for values of θ from $-360°$ to $720°$. (*Note:* To avoid confusion with x- and y-coordinates or the idea that t represents time and h represents height on the Ferris wheel, we are introducing new variables here.)

2. What is the amplitude of this function?

3. What is the period of this function? Why is the sine function periodic?

4. What are the θ-intercepts of the graph?

5. What values of θ make $\sin \theta$ a maximum? What values of θ make $\sin \theta$ a minimum?

6. If the equation $h = \sin t$ describes the "platform height function" for some Ferris wheel, what are the specifications of that Ferris wheel? (That is, what are its radius, its period, and the height of its center?) Indicate any ways in which it differs from the "standard" Ferris wheel described in *Homework 2: As the Ferris Wheel Turns*.

Sand Castles

Shelly loves to build elaborate sand castles at the beach. Her big problem is that her sand castles take a long time to build, and they often get swept away by the in-coming tide.

Shelly is planning another trip to the beach next week. She decides to pay attention to the tides so that she can plan her castle building and have as much time as possible.

The beach slopes gradually up from the ocean toward the parking lot. Shelly considers the waterline to be "high" if the water comes farther up the beach, leaving less sandy area visible. She considers the waterline to be "low" if there is more sandy area visible on the beach. Shelly likes to build as close to the water as possible because the damp sand is better for building.

According to Shelly's analysis, the level of the water on the beach for the day of her trip will fit this equation.

$$w(t) = 20 \sin 29t$$

In this equation, $w(t)$ represents how far the waterline is above or below its average position. The distance is measured in feet, and t represents the number of hours elapsed since midnight.

In the case shown in the accompanying diagram, the water has come up above its average position, and $w(t)$ is positive.

Continued on next page

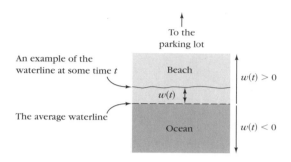

1. Graph the waterline function for a 24-hour period.

2. **a.** What is the highest up the beach (compared to its average position) that the waterline will be during the day? (This is called *high tide*.)

   b. What is the lowest that the waterline will be during the day? (This is called *low tide*.)

3. Suppose Shelly plans to build her castle right on the average waterline just as the water has moved below that line. How much time will she have to build her castle before the water returns and destroys her work?

4. Suppose Shelly wants to build her castle 10 feet below the average waterline. What is the maximum amount of time she can arrange to have to make her castle?

5. Suppose Shelly decides she needs only two hours to build and admire her castle. What is the lowest point on the beach where she can build it?

POW 1 Presentations

Mathematical Topics

- Applying the extended sine function in a new problem situation
- Using a recursion equation to understand the Tower of Hanoi problem

Outline of the Day

In Class

1. Discuss *Homework 6: Sand Castles*

- Use the graph to understand the behavior of the water level

2. Presentations of *POW 1: The Tower of Hanoi*

- Introduce the term **recursion**
- Have students explain the recursion equation in terms of the problem

At Home

Homework 7: More Beach Adventures

POW 2: Paving Patterns (due on Day 20)

1. Discussion of *Homework 6: Sand Castles*

You may want to give students a minute to check their work using graphing calculators before you begin the discussion. They might use the "trace" feature or a table to check for maximum and minimum points and to confirm results for Questions 2 through 5.

You can have volunteers present their results on each question.

- ## Question 1

  In the discussion of Question 1, have the presenter explain how he or she made the graph. For instance, the student may have realized that the first maximum occurs when $29t$ is equal to 90 (which gives $t \approx 3.1$) and that the water is at its average value again when $29t$ is equal to 180 (which gives $t \approx 6.2$). Help the class to recognize the connection between the water level graph and the graph of the "plain" sine function.

  Bring out that a 24-hour period will cover slightly less than two full periods for the water's motion. (You might ask at this point what the period of the function is, although that could also come out in connection with Question 3.)

  The graph might look like the diagram below, although students might also label their scales to indicate where the maximum and minimum points are, rather than show whole numbers of hours. (They also might use a 24-hour interval other than from $t = 0$ to $t = 24$.)

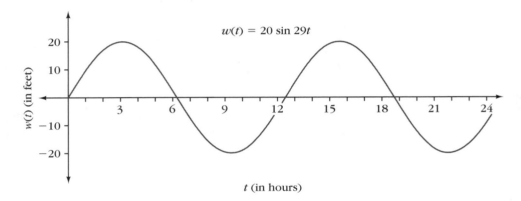

- ## Question 2

  For Question 2, students should see that the water level will get as high as 20 feet above and as low as 20 feet below the average waterline. Have them explain how they can see this at a glance from the function, and identify the number 20 as the amplitude of the function. (*Note:* The maximum value for $w(t)$ occurs at approximately $t = 3.1$ and 15.5, and the minimum at approximately $t = 9.3$ and 21.7. In other words, high tide is at about 3:06 a.m. and 3:30 p.m., and low tide is at about 9:18 a.m. and 9:42 p.m.)

- ## Question 3

  On Question 3, have the presenter give the specific times when the waterline is at its average level and identify these times on the graph of the function. Students should see that the question refers to the duration of the "lower half" of the curve. Be sure they identify the length of this time interval (which is about 6 hours and 12 minutes) as half the period of the function.

• Question 4

For Question 4, the presenter will probably have looked for solutions to the equation $w(t) = -10$, which simplifies to $\sin 29t = -0.5$. This is a good opportunity to review the idea of the inverse sine function and to discuss its limitations.

> *Note:* Students were introduced to the inverses of the *right-triangle* trigonometric functions in the Year 2 unit *Do Bees Build It Best?* The more general inverse sine function will be discussed tomorrow in connection with tonight's *Homework 7: More Beach Adventures.*

Students should see that their calculators give -30 as the value for $\sin^{-1}(-0.5)$. Discuss what this means, bringing out that the inverse sine function, as a function, can give only one of the angles whose sine is -0.5.

"How can you use the fact that $\sin(-30°) = -0.5?$"

Ask students how they can get the answer to Question 4 from the fact that $\sin(-30°) = -0.5$. They might use a graph, the Ferris wheel, or a coordinate system diagram to find other angles whose sine is -0.5.

In this problem, students are looking for two times when the waterline is at -10 feet: one (as the tide goes out) on the "down side" of the graph and one (as the tide comes back in) on the subsequent "up side" of the graph, as shown in the accompanying diagram. They should see that these two times could correspond to solutions of the equation $20 \sin 29t = -10$ with $29t$ representing angles in the third and fourth quadrants. This leads to the conditions $29t = 210°$ and $29t = 330°$, because both $\sin 210°$ and $\sin 330°$ are equal to -0.5. (As diagram indicates, Shelly could also use the next tide cycle, which gives the conditions $29t = 570°$ and $29t = 690°$.)

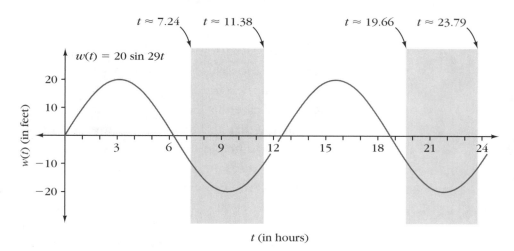

Shaded areas represent Shelly's options for the time interval in Question 4.

In other words, the water level reaches 10 feet below average about 7.24 hours after midnight (or at about 7:14 a.m.) and reaches that level again about 11.38 hours after midnight (or at about 11:23 a.m.). This gives Shelly about 4.14 hours to construct her sand castle. (The equations $29t = 570°$ and $29t = 690°$ give Shelly from about 7:39 p.m. to about 11:48 p.m., which is again a time interval of about 4.14 hours.)

- *Question 5*

  Question 5 is somewhat similar to Question 4. Again, students might use a graph or other diagram to get a sense of what's going on, perhaps seeing that Shelly needs to determine the water level one hour before low tide. They might use their knowledge about the period to see that the low tide first occurs about 9.3 hours after midnight, so Shelly should start about 8.3 hours after midnight. They then need to find $20 \sin (29 \cdot 8.3)$, which comes to approximately -17.5. In other words, if Shelly wants two hours in which to build her castle, the lowest she can go is 17.5 feet below the average waterline.

- *Sines without angles*

  If it hasn't yet come up, you may want to point out that the central equation in this problem involves the sine function, but the problem has nothing to do with angles. Tell students that the type of "rise and fall" motion that is involved in this problem and that is shown in the graph of the sine function occurs in many contexts that have nothing to do with angles.

2. Presentations of *POW 1: The Tower of Hanoi*

Let the three students chosen give their presentations, and then let other students share their ideas. Students should at least find the number of moves required for specific small numbers of discs.

- *Finding the minimum number*

  You can use the cases involving only a few discs to bring up the issue of how to show that these results are the smallest possible number of moves. For instance, by carefully examining the case of three discs, students should see the principle that it's "wasteful" to move the same disc on two consecutive moves. This principle shows that in the first three moves, one should get the two smallest discs onto another peg and that this can't be done in fewer than three moves. You can work further with this case to see why it's impossible to move all three discs in fewer than seven moves.

Students might use the information about specific cases in various ways to develop a general formula. If no general formulas are presented, you might have students put the results from individual cases into a table like this:

Number of discs	Number of moves needed
1	1
2	3
3	7
4	15

"What patterns or rules did you find?"

Have students describe patterns or rules that they see from this information. Two main observations are likely to come out of this table and out of students' work in generating it. If we use a_n to represent the minimum number of moves required if there are n discs, then the two key patterns can be written as follows:

- $a_n = 2^n - 1$
- $a_{n+1} = 2a_n + 1$

One proof of the first relationship involves the second relationship, together with an approach similar to mathematical induction. We suggest that you use your judgment about how much of the discussion we present here will be appropriate for your students.

- **The closed form: $a_n = 2^n - 1$**

    At least some of your students likely will see that the entries are each one less than a power of 2. In other words, the general formula for the number of moves required to move n discs is $2^n - 1$.

- **The recursive pattern: $a_{n+1} = 2a_n + 1$**

    Another pattern is that each successive entry is obtained by doubling the previous entry and adding 1.

"Why does this pattern hold true?"

Some students may have observed this pattern and seen why it works in the course of doing examples. If so, they can probably articulate fairly well why it works. If not, one way to help students get some insight into this is to have a student act out the process for, say, four discs, but to interrupt after three discs have been moved to another peg. Help them to see the process of moving four discs in three stages.

- Move three discs to the middle peg (7 moves)

- Move the biggest disc to the right peg (1 move)

- Move three discs from the middle peg to the right (7 moves)

(They should notice that they cannot move the fourth disc until the other three are all on the same one of the other two pegs.)

It is important to articulate that moving three discs from one peg to another is the same no matter which two pegs are involved. That's why the first and last stages in the three-stage process just described each take the same number of moves as solving the original puzzle for three discs. You simply have to be careful about which peg you move the first disc to.

To reinforce this idea, you might ask what the largest number of discs was for which anybody actually carried out the process. Suppose, for example, someone did eight discs, in 255 moves.

"How can you use the answer for eight discs to help you get the answer for nine discs?"

Ask how that information could be used to figure out the number of moves needed for nine discs. Students should see that they can move eight of the discs to the middle in 255 moves, use one move to move the largest disc to the right, and then use 255 more moves to move the pile of eight from the middle to the right. This gives a total of $255 + 1 + 255 = 511$ moves.

In other words, if we use a_n to represent the number of moves required to move an n-disc pile, then we see that this relationship holds:

$$a_{n+1} = a_n + 1 + a_n$$

(Students may simplify this to $a_{n+1} = 2a_n + 1$, but you might prefer to use the unsimplified equation because that reflects the sequence in which the groups of moves are done.)

Introduce the word **recursion** to describe the process by which each term in a sequence is described in terms of the preceding term or terms. The relationship $a_{n+1} = a_n + 1 + a_n$ (or $a_{n+1} = 2a_n + 1$) is called the **recursion equation.**

• Proving the closed formula

"Why does the closed formula work?"

If you have discussed both the recursion formula and the closed formula $a_n = 2^n - 1$, ask students if they have any explanation or proof for why the closed formula works. Point out that they can see that it works by example for the specific cases in the table, but that doesn't mean that it always works.

You can ask students to imagine that they know that the rule $2^n - 1$ works for a specific large value of n, such as $n = 20$. That is, they should suppose that they have verified somehow that moving 20 discs requires $2^{20} - 1$ moves.

"Can you find the number of moves for 21 discs without multiplying out 2^{20}?"

Ask them to find the number of moves for 21 discs *without multiplying out 2^{20}*. Tell them that they can use the recursive pattern.

"What do you get when you simplify $(2^{20} - 1) + 1 + (2^{20} - 1)$?"

They should see that 21 discs will require $(2^{20} - 1) + 1 + (2^{20} - 1)$ moves. Ask them to simplify this (again, without doing the arithmetic of finding 2^{20}).

They should see that the expression simplifies to $2 \cdot 2^{20} - 1$, which is the same as $2^{21} - 1$.

"Will this work for any number of discs?"

Ask whether this argument would work for any number of discs. Students should see that it does, and you can ask them to try to produce a general argument that if n discs can be transferred in $2^n - 1$ moves, then $n + 1$ discs can be transferred in $2^{n+1} - 1$ moves. As discussed earlier, the three-stage process gives the formula

$$a_{n+1} = a_n + 1 + a_n$$

so if n discs take $2^n - 1$ moves, then $n + 1$ discs take $(2^n - 1) + 1 + (2^n - 1)$ moves. But this simplifies to $2^{n+1} - 1$ moves.

In other words, this reasoning shows that if the formula $2^n - 1$ works for a particular number of discs, it also works when there is one more disc. If students get this far, tell them that this explanation is an example of a form of proof called **mathematical induction.** Tell them that this type of proof involves two elements: getting started (such as showing that the formula works in the case of only one disc) and going from one stage to the next (such as proving the recursion equation).

Note: This argument shows that n discs can be transferred in $2^n - 1$ moves. To show that this is the minimum possible number of moves, we need to argue that, essentially, in order to transfer a pile of $n + 1$ discs to the right, we must first move the first n of them to the middle, because this is the only way to move the last disc itself to the right. If students analyzed the case of three discs carefully, they can build on their reasoning.

As a side question, you can ask students to figure out which peg to move the first disc to in order to solve the puzzle in the minimum number of moves. They might see that it depends on whether n is odd or even.

• What about the monks?

Oh yes, let's not forget the monks. With 64 discs, the task would take $2^{64} - 1$ moves, or approximately $1.8 \cdot 10^{19}$. A series of computations shows that $1.8 \cdot 10^{19}$ seconds is about 580 billion years. Doing only 40 discs would take a mere 35,000 years. (Siddhartha Gautama, founder of Buddhism, lived from about 563 B.C.E. to 483 B.C.E. If the monks had begun this task when he was alive and moved one disc per second, they would now have moved the thirty-seventh disc and would be rebuilding the pile of the first 36 discs on top of it.)

Homework 7: More Beach Adventures

In this assignment, students continue their work with the situation of *Homework 6: Sand Castles* and do other work with the new extended definition of the sine function.

You may want to briefly review the use of the inverse sine function for this assignment.

POW 2: Paving Patterns

This POW will give students another opportunity to work with recursion. This POW is scheduled to be discussed on Day 20.

More Beach Adventures

1. After spending some of the day at the beach building sand castles, Shelly wants to take an evening walk with a friend along the shoreline.

Shelly knows that at one place along the shore, it is quite rocky. At that point, the rocks jut into the ocean so that in order to pass around them, a person has to walk along a path that is 14 feet below the average waterline.

Assume that Shelly and her friend don't want to get their feet wet. Therefore, they need to take their walk during the time when the waterline is 14 feet or more below the average waterline.

What is the time period during which they can take their walk?

(Recall that the position of the waterline over the course of the day is given by the equation $w(t) = 20 \sin 29t$, where the distance is measured in feet and t represents the number of hours elapsed since midnight.)

Continued on next page

2. Shelly often finds herself looking for numbers whose sine is a given value. This question asks you to do the same. Your solutions should all be between $-360°$ and $360°$. In Questions 2a and 2b, find exact values for θ. In Questions 2c and 2d, give θ to the nearest degree.

a. Find three values of θ, other than $15°$, such that $\sin \theta = \sin 15°$.

b. Find three values of θ such that $\sin \theta = -\sin 60°$.

c. Find three values of θ such that $\sin \theta = 0.5$.

d. Find three values of θ such that $\sin \theta = -0.71$.

POW 2

Paving Patterns

Al and Betty are hard at work. They are helping Al's family lay paving stones for a path along the side of their house.

The path is to be exactly 2 feet wide. Each paving stone is rectangular, with the dimensions 1 foot by 2 feet.

You might think this would be easy: simply lay one stone after another across the path. But there is more than one way to lay out the stones.

For example, a 3-foot section of the path could use any of the three arrangements shown here.

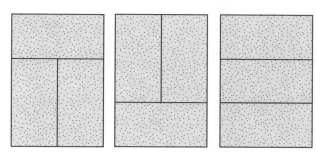

Important: These arrangements are all considered different, even though the first two are very similar.

So, of course, Al and Betty want to know how many different ways there are to lay out the stones. The path is 20 feet long altogether.

Al and Betty started to analyze this situation by using 1-inch-by-2-inch plastic tiles set out within a 2-inch-by-20-inch rectangle, but they soon were overwhelmed by all the possibilities.

Continued on next page

Can you help? You might want to start with shorter paths and look for patterns in the number of cases.

Note: You do not need to show the patterns themselves, except to explain your thinking. Your focus should be on *how many* patterns there are for a given length path.

Write-up

1. *Problem statement*

2. *Process*

3. *Results*

- Give the numerical results for any specific cases you studied.

- Give any general formulas that you found, even if you aren't sure of them.

- Give any explanations you found for your formulas.

4. *Evaluation*

5. *Self-assessment*

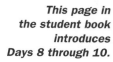

DAYS
8–10

Falling, Falling, Falling

This page in the student book introduces Days 8 through 10.

Dan Medina and Mike Rodricks add to their carefully kept notes a formula for the height of an object falling from rest.

The diver on the Ferris wheel doesn't simply go round and round. At some point, the assistant lets go and the diver begins his fall.

How long will the diver be in the air? You'll have to learn some principles of physics, as well as use some mathematics, to answer this question. The question is complicated by the fact that the diver does not fall at a constant speed.

Distance with Changing Speed

Mathematical Topics

- Defining the inverse sine function using principal values
- Expressing total distance traveled in terms of area under the graph of the speed function
- Developing a method for finding total distance traveled for situations involving constant acceleration

Outline of the Day

In Class

1. Discuss *Homework 7: More Beach Adventures*

- Bring out that defining the inverse sine function requires choosing principal values

2. Introduce the idea of using an area model to represent distance in terms of speed

3. *Distance with Changing Speed*

- Students find the total distance traveled under different circumstances

4. Discuss *Distance with Changing Speed*

- Establish that under constant acceleration, the average speed for a time interval is the average of the initial and final speeds for that interval

At Home

Homework 8: Acceleration Variations and a Sine Summary

Physics and Mathematics

As you will see, this unit requires students to understand and apply some important ideas from physics, especially principles involving falling objects and about the interaction of vertical and horizontal components of velocity. If your background in physics is not strong, you may feel a bit intimidated by this material.

What resources are available to you in understanding these ideas and helping your students to understand them? How can you enlist the aid of your school's science faculty? What suggestions do they have for explaining these principles?

1. Discussion of *Homework 7: More Beach Adventures*

Question 1 is somewhat similar to Question 4 of *Homework 6: Sand Castles.* You may want to give students a few minutes to compare ideas on Question 1, and then have a spade card student explain his or her solution.

The problem involves solving the equation $w(t) = -14$, which means finding values of t that fit the condition $20 \sin 29t = -14$. But students need to consider the periodicity of the function in order to find the two solutions for t that correspond to the evening hours. For instance, if they consider "evening" to mean a time between 6 p.m. and midnight, then they need t to be a value between 18 and 24.

By using the inverse sine function, students will come up with the "basic" solution to the equation. The equation $29t = \sin^{-1}\left(-\frac{14}{20}\right)$ yields $t \approx -1.53$, which corresponds to about an hour and a half before midnight *on the previous day.* There are several ways that students might find a solution for the evening of the day in question.

One approach is to use the condition $29t \approx -44.4$ to get other values for the expression $29t$. Using the graph of the function $w(t)$ (on the next page) may help in understanding what's happening here.

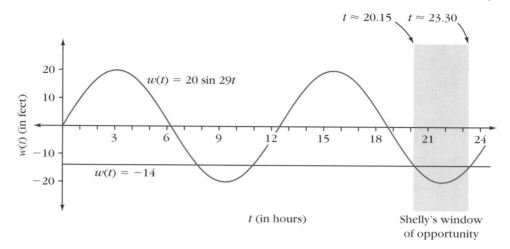

Given that the basic solution is $29t \approx -44.4$, students can use the graph to see that Shelly and her friend can pass by where the rocks jut into the water between $29t = 540° + 44.4°$ and $29t = 720° - 44.4°$. This gives the time interval from $t \approx 20.15$ and $t = 23.30$, which means from about 8:09 p.m. to about 11:18 p.m.

- ## Question 2

    Questions 2a through 2d involve a more abstract version of the issues in Question 1, although these questions are simpler because they involve merely $\sin \theta$ instead of $\sin 29\theta$. You might have several volunteers give an answer for each of the questions.

- ## Principal value of the sine function

    For Question 2d, many students will have used the "$\sin^{-1}$" key on their calculators. Point out that the calculator has to somehow choose one of many answers as its output for $\sin^{-1}(-0.71)$, and so it is important to standardize the process for doing this.

    You can let students explore for a few minutes to see how the calculator's "$\sin^{-1}$" key works. They should discover these facts.

    - If x is between 0 and 1, $\sin^{-1} x$ is between 0° and 90°.

    - If x is between 0 and -1, $\sin^{-1} x$ is between 0° and $-90°$.

    Of course, if the calculator is given a number with absolute value more than 1, it gives an error message when you use $\sin^{-1}$; see the next subsection, "The domain of inverse sine."

    > *Note:* With most scientific calculators, you enter the number (-0.71, for instance) and then push the "$\sin^{-1}$" key.
    >
    > With many graphing calculators, you do the reverse.

Tell students that although each of the answers to Question 2d is *an* inverse sine for −0.71, the value that the calculator gives (approximately −45.2°) is sometimes called the **principal value** for the inverse sine of −0.71. We write this using notation such as $\sin^{-1}(-0.71) \approx -45.2°$.

Bring out that the general definition of the inverse sine involves a *convention* for choosing these principal values.

> *Note:* Some textbooks introduce the notation "Sin⁻¹" (with a capital "S") to represent the principal value, and use "sin⁻¹" to be "multivalued." This is not standard mathematical notation, however.

You may want to point out the analogy between this distinction and a similar distinction with square roots. For example, the number 9 has two square roots, 3 and −3, but the symbol $\sqrt{9}$ represents only the number 3, which is sometimes called the *principal square root* of 9.

• The domain of inverse sine

"What's sin⁻¹ (2)?"

Ask students to find $\sin^{-1}(2)$ on their calculators. They should get some kind of error message.

Ask students to explain this. As a hint, ask them to express the problem in terms of the sine function (rather than in terms of inverse sine). They should be able to describe this as looking for a solution to the equation $\sin x = 2$. You may want to let them discuss this equation in their groups until someone can explain why this problem has no solution.

"What is the domain of the inverse sine function? What is its range?"

Use this example to review terminology by asking for the *domain* and *range* of the inverse sine function. Bring out that the domain is the interval from −1 to 1, because the equation $\sin x = c$ has a solution only if $-1 \le c \le 1$ or, in other words, because the range of the sine function is the interval from −1 to 1.

The range of $\sin^{-1}$ is the set of angles from −90° to 90°, inclusive, although this is the result of the conventions for principal values.

2. An Area Model for Distance

Remind students that one type of motion they will be considering will be that of the falling diver, whose speed changes as he falls. Tell them that because the speed is changing, the relationship among the variables of distance, speed, and time is more complex than if the speed were constant. Inform them that in order to understand this complex situation, they will develop a simple model, using a graph, for representing the distance a moving object (or person) travels in terms of its speed.

Begin by posing this straightforward question.

> *Suppose a person drives for 3 hours at a constant speed of 50 miles per hour. How far does the person go?*

All students need to do here is multiply the speed (50 miles per hour) by the time (3 hours) to get the distance (150 miles).

Have students make a graph showing speed as a function of time for this situation. Their graphs should look like the one shown here.

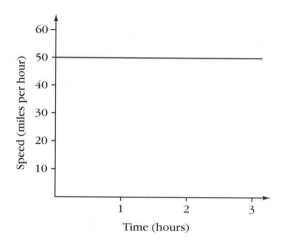

"How can you represent the distance geometrically?"

Then ask students how they might use this graph to get a geometrical representation of the distance the person traveled. As a hint, remind them that for constant speed, distance is simply the product of the speed and the time. As a further hint, remind them that the area of a rectangle is often a good model for multiplication.

These hints should lead the class to form a rectangle as in the next diagram and to see that the area of this rectangle gives the numerical value of the distance traveled. You may want to suggest that students subdivide the rectangle, as shown here, to indicate the distance covered each hour.

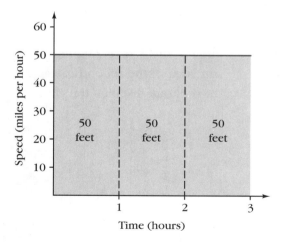

Tell the class that the next activity continues the use of this area model for finding total distance.

> *Comment:* It may seem strange to use an area model to represent a linear measurement, as the discussion here does. But because *distance* is the product of *rate* and *time,* it is appropriate to use a two-dimensional model for distance. (You may want to acknowledge this anomaly to students.)

3. *Distance with Changing Speed*

With the preceding introduction to the idea of an area model for finding distance, have groups work on the next activity, *Distance with Changing Speed*.

4. Discussion of *Distance with Changing Speed*

The purpose of Question 1 is to extend the use of the area model beyond the case of constant speed. For Question 1a, students should get a diagram something like this one showing speed as a function of time.

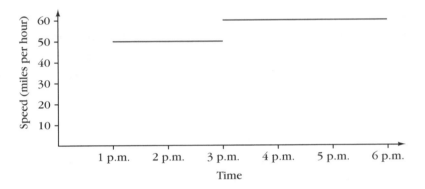

For Question 1b, students might draw in the rectangles shown in the next diagram to illustrate that the distance traveled from 1 p.m. to 3 p.m. is the area of the first rectangle and the distance traveled from 3 p.m. to 6 p.m. is the area of the second rectangle. Thus, the total area under the graph is equal to the total distance traveled.

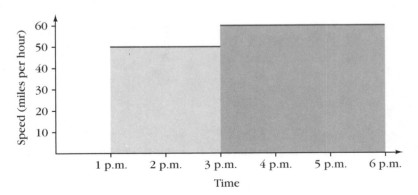

• Question 2

Question 2 applies the new area model to a situation in which speed is changing at a constant rate. For Question 2a, students should get a graph like the next one shown. (A larger version of this graph is included in Appendix B. You may want to use a transparency of this graph to aid in the discussion.)

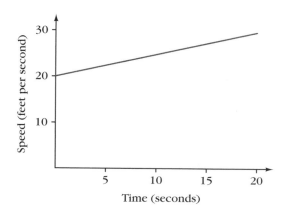

For Question 2b, we expect that students will take a purely intuitive approach, saying that because the speed increases at a constant rate from 20 feet per second to 30 feet per second, the average speed is simply 25 feet per second.

The main goal of this activity is to confirm this intuitive approach by using the area model. Based on the earlier examples, students should accept that the total distance traveled ought to be equal to the area under the graph, as shown here.

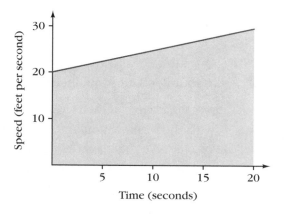

"How do you find the area of this shaded figure?"

Ask students how to find the area of this figure. They may recall (from the Year 2 unit *Do Bees Build It Best?*) that the figure is called a *trapezoid*, and they might even recall the formula for the area of a trapezoid. But don't get sidetracked by area formulas here.

"What rectangle would
have the same area,
using the same base?"

One intuitive approach is to ask what rectangle would have the same area, using the same base. Students should see that the rectangle shown in the next diagram has the same area as the trapezoid.

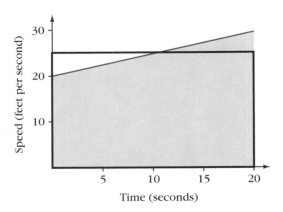

Students should also be able to see that the height of this rectangle is 25, which is the average of the heights of the two ends of the trapezoid, and that the area of the rectangle, and hence of the trapezoid, is 500. You may want to have them express the area in the form $\left(\frac{20 + 30}{2}\right) \cdot 20$ and review that this is an illustration of the general formula for the area of a trapezoid.

> _Comment:_ Trapezoids are commonly used in calculus as a way of approximating areas under curves.

"What does the
area mean in terms
of the runner?"

Ask what the area means in terms of the runner. Bring out that it means that the runner goes a total of 500 feet.

Finally, verify that this result is consistent with the intuitive answer to Question 2b. That is, if the runner averaged 25 feet per second, as in Question 2b, and went at this speed for 20 seconds, he would travel 500 feet, as found in Question 2c using area.

"What do we call the
rate at which speed
is changing?"

Ask what we call the rate at which speed is changing. (If no one comes up with the term, you might ask what we call it when a car speeds up.) If no one knows, give the class the term **acceleration.** Introduce the phrase **constant acceleration** to describe the situation in Question 2, in which the speed is changing at a constant rate.

> More precisely, the term _acceleration_ refers to the change in _velocity_ rather than the change in _speed._ The distinction between velocity and speed will be discussed on Day 24.
>
> For now, you can ignore that distinction because there is no change in the direction of the runner's motion.

- *"Averaging the endpoints"*

Bring out that the trapezoid approach will work for any situation of constant acceleration and that it gives a simple way to find the total distance traveled, even though the speed is not constant.

Ask students to sum up the principle for finding average speed in situations of constant acceleration. They should be able to articulate something like this, which you should post.

> **If an object is traveling with constant acceleration, then its average speed over any time interval is the average of its beginning speed and its final speed during that time interval.**

We will refer to this principle as the "averaging the endpoints" method for finding average speed. Emphasize to students that it applies only to a situation of constant acceleration.

You might also ask how this principle can be used to find the total distance traveled. Students should see that, as always, they can multiply the average speed by the length of the time interval to get the total distance.

Homework 8: Acceleration Variations and a Sine Summary

The first part of this assignment continues students' work with the area model for distance. The second part asks students to summarize their work in extending the sine function and to connect this new definition with the unit problem. The main purpose of Part II is to get them thinking about the issues as preparation for a class discussion.

You may want to give some students transparencies to prepare graphs for tomorrow's discussion of Part I of this assignment.

Important: If the class has not completed the discussion of *Distance with Changing Speed,* you should postpone Part I of this assignment.

Distance with Changing Speed

1. Curt drove from 1 p.m. to 3 p.m. at an average speed of 50 miles per hour and then drove from 3 p.m. to 6 p.m. at an average speed of 60 miles per hour.

a. Draw a graph showing Curt's speed as a function of time for the entire period from 1 p.m. to 6 p.m., treating his speed as constant for each of the two time periods—from 1 p.m. to 3 p.m. and from 3 p.m. to 6 p.m.

b. Describe how to use areas in this graph to represent the total distance he traveled.

2. Consider a runner who is going at a steady 20 feet per second. At exactly noon, he starts to increase his speed. His speed increases at a constant rate so that 20 seconds later, he is going 30 feet per second.

a. Graph the runner's speed as a function of time for this 20-second time interval.

b. Calculate his average speed for this 20-second interval.

c. Explain how to use area to find the total distance he runs during this 20-second interval.

Acceleration Variations and a Sine Summary

Part I: Acceleration Variations

In Question 2 of *Distance with Changing Speed,* you considered the case of a person running with constant acceleration. In other words, the runner's speed was increasing at a constant rate.

In that problem, the runner's speed went from 20 feet per second to 30 feet per second over a 20-second time interval. The accompanying diagram shows a graph of the speed as a function of time.

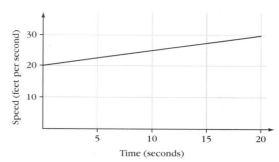

That problem illustrated an important principle.

If an object is traveling with constant acceleration, then its average speed over any time interval is the average of its beginning speed and its final speed during that time interval.

Continued on next page

According to this principle, the runner's average speed for the 20-second interval was exactly 25 feet per second, which is the average of 20 feet per second (the beginning speed) and 30 feet per second (the final speed).

Your task in Part I of this assignment is to describe three variations on this situation. In each case, the runner's speed should increase, as before, from 20 feet per second to 30 feet per second over the same 20-second time interval. But in your examples, *the runner's acceleration should not be constant*.

For each example, show a graph of the runner's speed in terms of time.

• Give an example in which the runner's average speed is *more than* the average of the beginning speed and the final speed.

• Give an example in which the runner's average speed is *less than* the average of the beginning speed and the final speed.

• Give an example in which the runner's average speed is *equal to* the average of the beginning speed and the final speed. (Remember that even with this example, the runner's acceleration should not be constant.)

Part II: A Sine Summary

The idea of extending the sine function to all angles — not merely acute angles — is an important concept. Your task in Part II of this assignment is to reflect on your work with the sine function. Your written work should include these things.

• A summary of what you have learned so far about this idea

• Any questions you still have about this extended sine function

• An explanation of how the extension of the sine function helps with the solution of the unit problem

DAY 9

Free Fall

Mathematical Topics

- Summarizing ideas about extending the sine function
- Recognizing that falling objects accelerate and learning that their acceleration is constant
- Using the "averaging the endpoints" method in connection with falling objects

Outline of the Day

In Class

1. Discuss *Homework 8: Acceleration Variations and a Sine Summary*

2. Discuss basic ideas about falling objects

- Review the "averaging the endpoints" method
- Connect this method to the unit problem through the principle that falling objects have constant acceleration

3. *Free Fall*

- Students apply the "averaging the endpoints" method to find a formula for the height of a falling object in terms of time
- The activity will be discussed on Day 10

At Home

Homework 9: Not So Spectacular

1. Discussion of *Homework 8: Acceleration Variations and a Sine Summary*

- *Part I: Acceleration Variations*

    Let a different volunteer share an example for each of the variations called for in Part I of the assignment. You might give presenters transparencies of the graph shown in the assignment and have them each sketch their graphs using the axes on the transparency. Like the graph in the assignment, the graphs students create should show the speed going from 20 feet per second to 30 feet per second. Have each presenter explain how he or she knows that the graph fits the given condition.

- *More about acceleration*

    Point out that the units for acceleration are somewhat complicated. Speed itself is measured here in feet per second, and acceleration measures how speed changes over time. In this example, the runner's speed increases by 10 feet per second over a 20-second interval, so it increases by 0.5 feet per second for each second that elapses. Tell students that we express this by saying that the acceleration is 0.5 feet per second per second.

    Emphasize the distinction between *speed,* which in this case tells the rate at which the runner's *position* is changing, and *acceleration,* which tells how fast the runner's *speed* is changing. Here, the *speed* is not constant (the runner's *position* is not changing at a constant rate), but the *acceleration* is constant (the runner's *speed* is changing at a constant rate).

- *Part II: A Sine Summary*

    Let students share their ideas with the class about the extension of the sine function and how it fits with the problem. Students should feel fairly comfortable at this stage of the unit with the idea that the expression $65 + 50 \sin 9t$ gives the height of the platform after t seconds (starting from the 3 o'clock position).

2. Falling Objects

"What does 'averaging the endpoints' have to do with the main unit problem?"

"What happens to an object as it falls freely?"

You might quickly review the principle of "averaging the endpoints." Then ask what this principle has to do with the main unit problem. If students don't see a connection, ask what happens to an object as it falls freely. Let students share their own ideas and experiences about what happens as objects fall.

If no one has a convincing argument that falling objects gain speed, then you may want to bring up a situation like this for them to consider.

> *Which would hurt more, a fall from your roof or a fall from your bed?*

(Although this example relates more directly to the force of impact than to speed, most students will attribute the added force of falling from a roof to moving faster at impact.)

Tell students that although our experience tells us that objects go faster and faster as they fall, physicists actually know much more precisely how falling objects behave. Specifically, from experimental data and from theoretical considerations, they know that falling objects have *constant acceleration*. You may want to post this statement, perhaps adjacent to the description of the "averaging the endpoints" method.

> **Falling objects have constant acceleration (under ideal circumstances). That is, the speed of a falling object changes at a constant rate.**

Discuss that the phrase "under ideal circumstances" means that there is no wind, air resistance, or other complicating factor to interfere with the object's fall. That is, the principle describes the behavior of *free-falling* objects. (This assumption is mentioned in the next activity.) You might also discuss the fact that this assumption is reasonable for some types of objects (such as rocks) and not for others (such as feathers).

> *Note:* The activity *Look Out Below!* (Day 21) involves a falling pillow, and a different rate of acceleration will be used in that problem.

3. *Free Fall*

Have students read the introduction to the activity *Free Fall* and go over the details in the section "Starting from Rest." Then have them work on the questions. If necessary, suggest that for Question 1, they use the "averaging the endpoints" method.

Note: Question 5 explicitly states that the diver falls "from rest." If the issue of the effect of the Ferris wheel's motion has come up before (see the subsection "For teachers: The diver's initial motion" on Day 2), you may want to remind students that they are assuming for now, in their work on the central unit problem, that the diver falls from rest. You can tell them that they will drop this simplification later in the unit (see the section "But There's More to the Problem!" on Day 16).

Homework 9: Not So Spectacular

Tonight's assignment resembles *Homework 3: A Clear View* but continues from the work in that problem to deal with the idea of the periodicity of the Ferris wheel's motion. This assignment also puts some of the ideas from *Homework 7: More Beach Adventures* in the context of the Ferris wheel.

CLASSWORK

Free Fall

As long as people can remember, objects have been falling. But it wasn't until the sixteenth and seventeenth centuries that scientists fully understood the physics and mathematics of falling objects.

The Italian physicist Galileo Galilei (1564–1624) is one of those credited with figuring out the laws of gravitational fall based on experiments. The English physicist Isaac Newton (1642–1727) developed a broader theory of gravitation to explain Galileo's observations.

Free-Falling Objects

Using both experiments and theoretical analysis, physicists have confirmed this principle.

Falling objects have constant acceleration.

This principle assumes that there is no air resistance or other complicating factor to interfere with the object's fall. That is, the principle describes the behavior of *free-falling* objects. In this unit, you should assume, unless told otherwise, that all falling objects are falling freely.

The analysis by physicists is even more precise than this broad principle.

The instantaneous speed of a freely falling object increases approximately 32 feet per second for each second of its fall.

Continued on next page

Starting from Rest

The simplest case occurs when the object starts from rest, that is, when its speed is zero when $t = 0$. In this case, the object's instantaneous speed after 1 second is 32 feet per second; after 2 seconds, its instantaneous speed is 64 feet per second; and so on.

From Acceleration to Distance

Your task in this activity is to use the principles just stated to express the distance an object falls in terms of the amount of time it has been falling. You should assume that the object is dropped from rest and falls freely.

1. **a.** How fast is the object going at $t = 5$?

   b. How far does the object fall in its first 5 seconds?

2. Generalize your work from Question 1 to develop a formula for how far the object falls in its first t seconds.

3. Suppose the object starts from a height of h feet. What is its height after t seconds? (Assume that the object has not yet reached the ground.)

4. Use your result from Question 3 to find an expression in terms of h for the amount of time it would take for the object to reach the ground.

Now apply your work to a simple version of the circus act.

5. Suppose the platform is fixed at 90 feet above the ground, the diver falls freely from rest, and the level of the water in the tub is 8 feet above the ground. How long will it take for the diver to reach the water?

Not So Spectacular

The circus owner decided that to save money, he would fill in for the diver from time to time.

This did not end up being a very good idea because the owner was not an experienced diver, so he could not safely fall large distances. In fact, he refused to be dropped from more than 25 feet above the ground. He also insisted that there be a huge tub of water under him at all times.

Your task is to find all possible times when the owner will be 25 feet from the ground. You may want to describe the complete set of possibilities by writing an algebraic expression for t.

Reminder: Here are the basic facts about the Ferris wheel.

• The radius is 50 feet.

• The center of the Ferris wheel is 65 feet off the ground.

• The Ferris wheel turns counterclockwise at a constant angular speed, with a period of 40 seconds.

• The Ferris wheel is at the 3 o'clock position when it starts moving.

Students find a formula for the height of a falling object as a function of time.

The Height of a Falling Object

Mathematical Topics

- Working with the fact that there are many angles with the same sine
- Finding a formula for the height of an object falling from rest as a function of time

---

Outline of the Day

In Class

1. Discuss *Homework 9: Not So Spectacular*

- Develop a general expression for the times at which the diver will be at a given height

2. Discuss *Free Fall* (from Day 9)

- Develop a formula for the height of an object falling from rest as a function of time
- Use the height formula to develop a formula for the time it takes for an object to fall *h* feet

At Home

Homework 10: A Practice Jump

---

1. Discussion of *Homework 9: Not So Spectacular*

We suggest that you give students a few minutes to compare ideas, and perhaps select a student at random to present a solution for the homework.

You may want to let several students contribute as you work toward the general solution. The discussion here is intended as a guide for the homework discussion, but the ideas should come from the class.

Up to a point, this problem is essentially the same as *Homework 3: A Clear View*. In both problems, students must find the points in the cycle of the Ferris wheel where a person is a given height off the ground. The difference in this problem is the focus on obtaining the general solution. You may want to have several students contribute as the class works toward that general solution. Use this discussion to review the ideas about the inverse sine function and principle values discussed in connection with *Homework 7: More Beach Adventures*.

• *Initial solutions*

Encourage presenters to use a diagram to explain their work. The diagram here shows the two positions at which the owner would be 25 feet off the ground and gives details about one of these locations. This shows that the angle θ must satisfy the condition $\sin \theta = \frac{40}{50}$, so $\theta = \sin^{-1} 0.8$, which is approximately $53.1°$. This means that the large angle shown as $9t$ would be equal to $360° - 53.1° = 306.9°$. In other words, the Ferris wheel could have been turning for about $306.9 \div 9 = 34.1$ seconds.

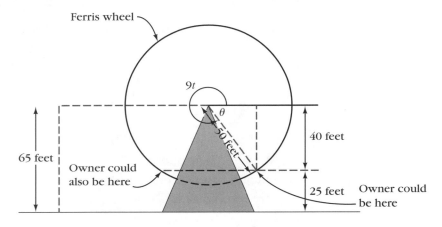

For the other location at a height of 25 feet, we have

$$9t \approx 180° + 53.1° = 233.1°,$$

which means $t \approx 233.1 \div 9 = 25.9$ seconds.

To this point, the problem is basically a repetition of the ideas in *Homework 3: A Clear View*.

• *Generalizing the solutions*

The new aspect of this problem is that each of these two locations actually leads to an infinite list of possible answers to the question of how long the Ferris wheel has been turning, because the Ferris wheel may have gone around several times before reaching the given point. (If students did not realize that the homework problem called for this additional stage, let them work on it in groups, trying to find the general solutions.)

Two main approaches are likely to come about. Here are descriptions of these two approaches for the position at the lower right of the diagram:

- Find different possibilities for the total angle 9*t*, by adding multiples of 360° to the basic angle of 306.9°, and then divide each total angle by 9 (degrees per second).

- Find the time of 34.1 seconds for the basic angle of 306.9° and then add multiples of 40 seconds to that time.

In either case, one gets 34.1 seconds, 74.1 seconds, 114.1 seconds, and so on as the sequence of possible times. Students should be able, perhaps with some prodding, to express this general solution as something like 34.1 + 40*n*, where *n* can be any integer.

For the position at the lower left of the diagram, the same steps apply, but the general solution is 25.9 + 40*n* seconds.

Bring out that all of these values satisfy the equation sin 9*t* = −0.8, but only one is based on the principal value of the inverse sine function.

2. Discussion of *Free Fall*

Let one or two heart card students present their analysis for Question 1. Question 1a should be straightforward. That is, if students understood the introduction to the activity, they should see that the speed at *t* = 5 is simply 5 · 32 = 160 ft/sec.

To find out how far an object falls in 5 seconds, students should reason that the object's average speed for that interval is equal to the average of its instantaneous speeds at the endpoints of the interval. These endpoints are *t* = 0 and *t* = 5. The information in the activity tells them that the instantaneous speed at *t* = 0 is 0 ft/sec and the instantaneous speed at *t* = 5 is 160 ft/sec.

The average of 0 and 160 is 80, so the object has an average speed over the 5-second interval of 80 ft/sec. Therefore, the object falls 5 · 80 = 400 feet during this interval.

- *Question 2*

  The key element in this activity is for students to generalize the reasoning from Question 1 to develop the general formula asked for in Question 2.

The approach we expect is for students to see that the instantaneous speed at the end of t seconds is $32t$ and the instantaneous speed at the start is 0. Thus, the average speed over the first t seconds is $16t$ ft/sec. Students can then multiply this average speed by the length of the time interval, which is t seconds, to get a total distance traveled of $16t^2$ feet.

Post this conclusion, because it will play a critical role throughout the rest of the unit.

If an object falls freely from rest, it will fall $16t^2$ feet in its first t seconds.

Note: Physicists often use the term "displacement" and the letter s for the distance an object has traveled. Some students may have seen the formula $s = 16t^2$ in a physics class to describe the displacement of an object falling from rest.

• Questions 3 and 4

For Question 3, students simply need to subtract $16t^2$ from the initial height h to get the formula $h - 16t^2$ for the object's height after t seconds. You can add this additional conclusion to the statement just posted.

If the object's initial height is h feet, then its height after t seconds is $h - 16t^2$ feet.

Question 4 may seem straightforward once this formula has been found, but be sure to go over the transition carefully. For some students, it may be a substantial step from the idea of "reaching the ground" to the step of setting $h - 16t^2$ equal to 0. Help them to understand that answering Question 4 is equivalent to solving the equation $h - 16t^2 = 0$ for t in terms of h. Have a volunteer go over the details of solving the equation to get $t = \sqrt{\frac{h}{16}}$ seconds.

Once students have developed this additional generalization, you should post it along with the previous formula. This new statement might say something like this.

If an object falls freely from rest, it will take $\sqrt{\frac{h}{16}}$ seconds for the object to fall h feet.

(This formula is included in tonight's homework.)

• Question 5

Question 5 provides an important variation, in which students need to see that the diver is actually falling 82 feet. They might set this up through the equation $90 - 16t^2 = 8$, or they might simply set $16t^2$ equal to 82. In either case, they should get the expression $\sqrt{\frac{82}{16}}$, which means that it takes approximately 2.26 seconds for the diver to fall to the water level.

• *The number 32 is an approximation*

At some point, bring out that the number 32, which appears in *Free Fall*, is a numerical approximation based on experiments. (The activity does say "approximately" in giving the value of 32 feet per second for each second of the object's fall, but students may overlook this.)

> *Note:* The actual rate of acceleration depends on the force of gravity, which varies slightly from one place on earth to another and which is lower at higher altitudes.

Also point out that this number is specific to the use of feet as the unit of length. For example, if we instead measure length in meters, then we use approximately 9.8 instead of 32 (because 32 feet is about 9.8 meters).

Homework 10: A Practice Jump

> Tonight's homework asks students to combine their new formula for falling time with their earlier work on the height of the platform at a given position and to generalize this in terms of the amount of time the Ferris wheel has been turning.

"I was impressed with how well High Dive *meshed with the work the students were doing in physics. It was a great opportunity for some cross-curricular work. We had many discussions about where the students first learned the physics of motion; was it from their physics teacher or their IMP teacher?"*

IMP teacher Dave Robathan

A Practice Jump

After some not-so-high practice dives by the circus owner, the circus decided to do a practice run of the show with the diver himself. But they decided to set it up so that they would not have to worry about a moving cart.

Instead, the cart containing the tub of water was placed directly under the Ferris wheel's 11 o'clock position. As usual, the platform passed the 3 o'clock position at $t = 0$.

1. How many seconds will it take for the platform to reach the 11 o'clock position?

2. What is the diver's height off the ground when he is at the 11 o'clock position?

One purpose of this practice run was to see how long it would take for the diver to fall into the water. You should be able to predict this, based on the formula that an object falling freely from rest takes $\sqrt{\frac{h}{16}}$ seconds to fall h feet. (Assume that the diver is falling freely from rest.)

3. How long will it take from the time the diver is released until he hits the water? (Don't forget that the water level in the cart is 8 feet above the ground.)

4. More generally, suppose the assistant lets go W seconds after the Ferris wheel starts turning. (Here, W stands for "wheel time.") Assuming that the cart is in the right place, how long will the diver be in the air before he hits the water?

DAYS 11-13

Moving Left and Right

This page in the student book introduces Days 11 through 13.

Ken Hoffman and Jeff Tung address the fact that the turning Ferris wheel also involves horizontal movement of the diver's platform.

Thus far, you have mostly been considering the platform's position and the diver's motion in the vertical dimension. But as the Ferris wheel turns, the platform is also moving to the left or right, and the cart is moving steadily to the right (once it gets started).

The key to a successful dive is to have the cart at the right place at the right time, which means it's now time for you to consider the horizontal dimension of the Ferris wheel problem.

DAY 11 Moving Left and Right

Mathematical Topics

- Combining principles about falling objects with the general sine function in connection with the unit problem
- Examining the horizontal dimension of the unit problem

Students set up a coordinate axis to study movement to the left and the right.

Outline of the Day

In Class

1. Form new random groups
2. Discuss *Homework 10: A Practice Jump*
   - Develop a function for the diver's falling time as a function of how long the platform has been turning
3. *Cart Travel Time*
   - Students determine how long the cart travels until the diver reaches the water level
4. Discuss *Cart Travel Time*
   - Identify the two parts of the cart's travel time and post the conclusion
5. Introduce the horizontal axis of the Ferris wheel coordinate system

At Home

Homework 11: Where Does He Land?

1. Forming New Groups

This is an excellent time to form new random groups. Follow the procedure described in the IMP *Teaching Handbook,* and record the members of each group and the suit for each student.

2. Discussion of *Homework 10: A Practice Jump*

Presentations of this assignment will be a good indicator of how well students understood yesterday's work. Let different volunteers present different parts of the problem. As usual, you should ask for alternate approaches after the presentations.

• Questions 1 and 2

One way to find the "let go" time (Question 1) is for students to see that the 11 o'clock position is one-third $\left(\frac{4}{12}\right)$ of the way around so it should take one-third of the time of a full rotation to get there. Another approach is to find the angle of rotation from the 3 o'clock position to the 11 o'clock position (120°) and divide this by the angular speed of 9 degrees per second.

In either case, students should see that the assistant should let go after about 13.33 seconds (or that plus any multiple of 40 seconds—you need not get distracted here by this extra detail).

To solve Question 2, students can substitute the time just found, about 13.33 seconds, into the general height formula, $h = 65 + 50 \sin 9t$. This will give them the height of 108.3 feet. They can also use the angle directly, getting the height from the expression $65 + 50 \sin 120°$.

• Question 3

For Question 3, students need to subtract 8 feet from the height of 108.3 feet (found in Question 2) to get the distance the diver will fall before he hits the water in the cart. That is, the diver falls 100.3 feet.

To complete Question 3, they can substitute the value $h = 100.3$ into the expression $\sqrt{\frac{h}{16}}$. They should get that the diver's falling time is about 2.50 seconds.

Note: Once again, we remind you that treating the situation as if the diver is falling from rest is a simplification that will be eliminated later in the unit.

• Question 4

For Question 4, students need to combine the general formula for the diver's height after t seconds on the Ferris wheel with the formula from *Free Fall* for the time required to fall h feet. Essentially, this is merely substitution, which may sound simple enough, but it is difficult for some students to make this transition. So take it slow and be sure that presenters explain themselves clearly.

You may want to refer repeatedly to the specifics of Question 3 as this generalization is developed. Here are the key steps of the generalization.

- After W seconds, the diver is at a height of $65 + 50 \sin 9W$ feet.

- The diver needs to fall 8 feet less than this, for a total falling distance of $57 + 50 \sin 9W$ feet.

- For an object at rest to fall $57 + 50 \sin 9W$ feet takes $\sqrt{\frac{57 + 50 \sin 9W}{16}}$ seconds.

• *Falling time is a function of turning time*

Bring out that students are expressing the time the diver spends in the air in terms of the amount of time the Ferris wheel turns before the assistant lets go. You might do this through a series of questions as outlined here.

"What does the diver's falling time depend on?"

• Begin by asking students what the diver's falling time depends on. They will probably say that it depends on his height when the assistant lets go.

"What does this height depend on?"

• Then ask what that height depends on. They will probably say it depends on what the diver's position is when the assistant lets go.

"What does this position depend on?"

• Finally, ask what that position depends on. They should see that it depends on how long the Ferris wheel is turning before the assistant lets go.

Students may shortcut one or more of these steps. For example, they may say that the falling time depends on where the diver is, or even that it depends on how long the Ferris wheel is turning. The important thing is that they trace the falling time back to the amount of time the Ferris wheel turns before the assistant lets go.

"What are you trying to find in the main unit problem?"

Then ask students what they are trying to find in the main unit problem. They should see that they are being asked to find out how long the assistant should let the Ferris wheel turn before letting go.

In other words, what they are looking for is what was called W in Question 4 of last night's homework and in the formula they found for that question. That is, they have just found a formula for the diver's falling time in terms of the variable that they are trying to find in the unit problem.

This can be summarized as follows:

If the Ferris wheel passes the 3 o'clock position at $t = 0$ and turns for W seconds more before the assistant lets go, then the diver will fall for

$$\sqrt{\frac{57 + 50 \sin 9W}{16}}$$

seconds before reaching the level of the water in the cart.

Post this conclusion prominently. It synthesizes two major ideas—the time required for a free-falling object to fall a given distance and the diver's height at the time of release—into a single important formula. You may want to add it on to the poster for the height of the platform after W seconds (see Day 5).

Because this expression for falling time is so complex, you may find it helpful to use a single letter to represent it. We will use F ("falling time"), so F is given by the equation

$$F = \sqrt{\frac{57 + 50 \sin 9W}{16}}$$

If you use this abbreviation, be sure students keep in mind that *F* is a function of *W*. You may want to have several students express in their own words what *F* represents.

> *Note:* Students will develop a different expression for *F* when they consider the effect of the turning of the Ferris wheel on the diver's path of fall. That aspect of the unit begins with *Homework 20: Initial Motion from the Ferris Wheel*.

3. *Cart Travel Time*

In the next activity, *Cart Travel Time,* students need to use their conclusion about how long the diver is falling to reach a conclusion about the amount of time the *cart* is traveling. The key to this brief activity is to look at the cart's travel time in two parts.

- The time between when the platform passes the 3 o'clock position and when the diver is released from the platform

- The time while the diver is falling from the platform

> In Part I of *Homework 12: Carts and Periodic Problems,* students will find the cart's *position* at the moment when the diver reaches the water level.

4. Discussion of *Cart Travel Time*

Only a brief presentation should be needed here, because students already have expressions for each of the two parts of the cart's travel time.

- The cart travels for *W* seconds from when the platform passes the 3 o'clock position until the diver is released.

- The cart travels for *F* seconds while the diver is falling $\left(\text{where } F \text{ is given in terms of } W \text{ by the expression } F = \sqrt{\frac{57 + 50 \sin 9W}{16}}\right)$.

Post a summary of the conclusion from this activity.

If the cart begins moving when the Ferris wheel passes the 3 o'clock position, and the diver is released *W* seconds later, then the cart will travel for *W* + *F* seconds before the diver reaches the level of the water in the cart, where *F* is given by the equation

$$F = \sqrt{\frac{57 + 50 \sin 9W}{16}}$$

5. The Horizontal Dimension

Bring out that most of students' work on the unit problem has focused on vertical motion—the height of the platform as it turns and the falling motion of the diver.

"What will determine the success or failure of the circus act?"

Ask students what will determine the success or failure of the circus act. Help them to see that they must look at the horizontal dimension as well. That is, the cart needs to be in the right place along its horizontal path when the diver reaches the water level.

Introduce the horizontal axis as shown in the accompanying diagram, in which the center of the base of the Ferris wheel represents zero and in which positions to the right of the Ferris wheel are considered to have positive x-coordinates. (Distances are measured in feet.) Tell students that we will identify an object's horizontal position within this system using its x-coordinate.

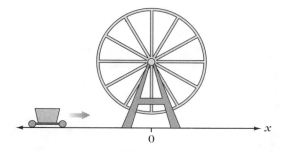

"What is the cart's x-coordinate at the start of the circus act?"

To clarify this system, ask students what the cart's x-coordinate is at the start of the circus act. They should see that the cart's initial x-coordinate is -240, because the cart is 240 feet to the left of the center of the base of the Ferris wheel.

"What about the platform?"

Similarly, ask about the platform's x-coordinate at the start of the circus act. Students should see that the platform's initial x-coordinate is 50, because the platform starts at the 3 o'clock position and the radius of the Ferris wheel is 50 feet.

Homework 11: Where Does He Land?

This assignment is like a combination of *Homework 2: As the Ferris Wheel Turns* and *Homework 4: Graphing the Ferris Wheel,* except that it deals with horizontal instead of vertical position.

Cart Travel Time

Thus far in the unit, you've focused mainly on the position and motion of the diver. So where's the cart of water in all this?

Recall that the cart starts moving when the Ferris wheel passes the 3 o'clock position. The goal is to have the cart be in the correct position when the diver reaches the level of the water in the cart. In this activity, you will consider only the cart's *travel time*.

Suppose the assistant lets go of the diver W seconds after the Ferris wheel passes the 3 o'clock position. Write an expression in terms of W for the length of time the cart will have traveled from the moment it starts until the moment when the diver reaches the level of the water.

Where Does He Land?

Earlier in the unit, you found that the expression $65 + 50 \sin 9t$ gives the diver's height off the ground while he is still on the platform. But what about the diver's *horizontal* position? This will be crucial in determining whether he lands in the tub of water on the moving cart.

To describe the diver's horizontal position, we will use a horizontal coordinate system as shown below, in which an object's x-coordinate is based on its distance (in feet) to the right or left of the center of the Ferris wheel, with objects to the right of the Ferris wheel considered as having positive x-coordinates.

For instance, in this coordinate system, the platform and diver have an x-coordinate of 50 when the platform passes the 3 o'clock position. Similarly, the cart starts its motion with an x-coordinate of -240, because it is initially 240 feet to the left of the center of the base of the Ferris wheel.

As usual, assume that the platform passes the 3 o'clock position at $t = 0$. Recall also that the Ferris wheel turns counterclockwise at a constant rate with a period of 40 seconds. Further, assume that the diver falls straight down once he is released.

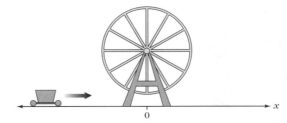

Continued on next page

1. Where will the diver land if he is released at each of these times?

    a. at $t = 3$

    b. at $t = 7$

    c. at $t = 12$

    d. at $t = 26$

    e. at $t = 37$

2. Sketch a graph giving the platform's x-coordinate as a function of t. Your graph should cover two complete turns of the Ferris wheel—that is, from $t = 0$ to $t = 80$.

    Note: Although the platform's x-coordinate represents its horizontal position, in this context x is a function of t. That is, t is the independent variable and x is the dependent variable. Therefore, the value of t should be shown on the horizontal axis of your graph, and the value of x on the vertical axis.

DAY 12

First-Quadrant Platform

Mathematical Topics

- Finding the platform's x-coordinate for specific cases
- Generalizing about the platform's x-coordinate for the first quadrant

Students find a formula for the x-coordinate of the platform for the first quadrant.

Outline of the Day

In Class

1. Discuss *Homework 11: Where Does He Land?*

- Use the cosine function in different quadrants to find the diver's x-coordinate as he falls

2. *First-Quadrant Platform*

- Students develop a general formula for the platform's x-coordinate in the first quadrant

3. Discuss *First-Quadrant Platform*

- Post the first-quadrant formula

At Home

Homework 12: Carts and Periodic Problems

Special Materials Needed

- A transparency of the blank coordinate system for showing the graph of the platform's horizontal position (see Appendix B)

1. Discussion of *Homework 11: Where Does He Land?*

For Question 1, let several students present their results for specific values of W, using diagrams to explain their answers. As these presentations are discussed, bring out that the x-coordinate of the diver's landing position is the same as the x-coordinate of the platform at the moment the diver is released.

For instance, for $t = 3$ (Question 1a), the Ferris wheel has turned 27°, and the presenter might give a diagram like the one shown here.

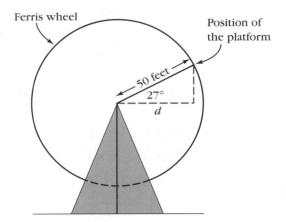

This gives $\cos 27° = \frac{d}{50}$, so the platform's x-coordinate is $50 \cos 27°$, or approximately 44.55. This means that while the diver is falling, his x-coordinate is also approximately 44.55. (*Comment:* Students may simply say that the diver lands about 44.55 feet to the right of the center of the Ferris wheel. Help them make the transition from this description to the use of the coordinate terminology.)

For the case $t = 12$ (Question 1c), the platform is in the second quadrant when the diver is released, and students are likely to express their common x-coordinate as $-50 \cos 72°$, as illustrated in the next diagram. It is important to bring out that although the segment labeled d has length $50 \cos 72°$ $\left(\text{because } \cos 72° = \frac{d}{50}\right)$, the x-coordinate must be negative. That is, we have $x = -50 \cos 72°$.

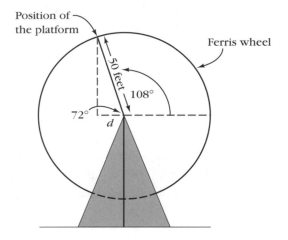

Similarly, for $t = 26$, students are likely to get $x = -50 \cos 54°$, while for $t = 37$, they will probably get $x = 50 \cos 27°$.

Note: Students might express these answers using the sine function instead, but ultimately, the goal is to get a general formula in terms of cosine.

• ## Question 2

.You can use a transparency of the blank coordinate system for this assignment provided in Appendix B, or you can have students develop the scales for the axes themselves. In either case, you or the students can plot individual points as they are suggested. Be sure to get a variety of points from $t = 0$ through $t = 80$. The graph should look roughly like the one shown here.

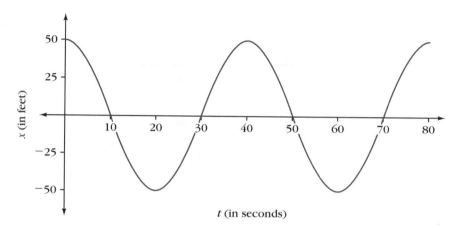

There are two main observations to make about this graph.

- This graph makes sense in terms of the Ferris wheel problem. For example, it is periodic with period 40, and it shows the platform being farthest to the right at $t = 0, 40$, and 80, farthest to the left at $t = 20$ and 60, and having a zero x-coordinate at $t = 10, 30, 50$, and 70.

- This graph is similar to the graph of vertical position that students made in *Homework 4: Graphing the Ferris Wheel*, with two key differences:

a) This graph has its maximum at $t = 0$ seconds, while the graph of the platform's height had its maximum at $t = 10$ seconds.

b) This graph is "balanced" around the horizontal axis, while the graph of the platform's height was "balanced" around the height of 65 feet.

2. First-Quadrant Platform

In this assignment, students will develop a general formula for the platform's x-coordinate for the first quadrant (as they did for the platform's height in *At Certain Points in Time* on Day 3). They will then use that formula to motivate the general definition of the cosine function. No whole-class introduction is needed, and probably only a brief discussion will be required.

3. Discussion of *First-Quadrant Platform*

Let a diamond card student present the group's result. The class should be able to see that the platform's x-coordinate is given by the equation $x = 50 \cos 9t$.

Homework 12: Carts and Periodic Problems

In Part I of this assignment, students need to combine their earlier formula for the cart's travel time with the newly introduced horizontal coordinate system. Part II provides another opportunity for students to think about periodic functions and periodic motion.

"Homework 13: Planning for Formulas *asks students, among other things, to 'put the pieces together' with the four formulas they have developed and to explain each formula clearly. So, when the students came into class with homework, I assigned a formula to each group, gave them butcher paper, and asked them to prepare a poster on their formulas. After twenty minutes, we posted the papers. It was wonderful! There was so much discussion going on as students read posters, added comments or thoughts, and, in some cases, made suggestions for corrections. I heard several students comment about how they now understood what was being taught. When the class sat down to begin* Moving Cart, Turning Ferris Wheel, *all were really in sync with the problem.*"

IMP teacher Cathy Martin

First-Quadrant Platform

In Question 1 of *Homework 11: Where Does He Land?*, you found the *x*-coordinate of the diver's landing position for a variety of specific cases.

At the moment when the diver is released, his *x*-coordinate is the same as the *x*-coordinate of the platform, and in Question 2 of that assignment, you sketched a graph of the platform's *x*-coordinate as a function of *t*.

Now focus specifically on values of *t* between 0 and 10, so that the platform is still in the first quadrant. Develop an equation that gives the platform's *x*-coordinate in terms of *t*.

Carts and Periodic Problems

Part I: Where's the Cart?

In *Homework 11: Where Does He Land?*, you looked at the horizontal coordinate of the diver as he falls. You also need to know where the cart is while the diver is falling and, especially, where it is when the diver reaches the water.

Recall that the cart begins 240 feet to the left of the base of the Ferris wheel, so its x-coordinate at $t = 0$ is -240. The cart moves to the right at 15 feet per second and begins moving at that speed at $t = 0$.

Based on this information, find the cart's x-coordinate at the moment when the diver reaches the water level.

Part II: Periodic Problems

In this unit, you have seen that the height of a platform on a Ferris wheel represents a periodic function. You have encountered periodic functions before. For instance, the swinging of a pendulum is periodic motion, and the bob's

distance from the center line is a periodic function of time (assuming that the pendulum isn't slowing down).

1. Describe three other situations that you believe are periodic. For each example, explain what is repeating and give the period for the repetition.

2. Sketch graphs of at least two of the periodic situations you described in Question 1.

DAY 13

The General Cosine Function

Mathematical Topics

- Considering examples of periodic behavior
- Finding a formula for the position of the cart
- Extending the cosine function to be defined for all angles
- Relating the general definition of the cosine function to the platform's *x*-coordinate

Outline of the Day

In Class

1. Discuss *Homework 12: Carts and Periodic Problems*
2. *Generalizing the Platform*
3. Discuss *Generalizing the Platform*
   - Emphasize that there is only one way to extend the cosine function that will allow the first-quadrant formula to work in all quadrants
4. Formally define the cosine function
   - Have students graph the function

5. Connect the cosine function to the Ferris wheel problem
   - Bring out that students now have general formulas for the diver's vertical and horizontal positions as functions of time

At Home

Homework 13: Planning for Formulas

1. Discussion of *Homework 12: Carts and Periodic Problems*

- ## *Part I: Where's the Cart?*

Have a volunteer present Part I. On Day 11, students saw that from $t = 0$ until the diver reaches the water level, the cart travels for $W + F$ seconds (where F represents the diver's falling time and is given by the expression $\sqrt{\frac{57 + 50 \sin 9W}{16}}$). Students need to combine that information with the facts about the cart's speed and initial position to get that the cart's x-coordinate when the diver reaches the water level is $-240 + 15(W + F)$.

Of course, you and the class will want to post this conclusion:

Suppose the diver is dropped after W seconds on the Ferris wheel (starting from the 3 o'clock position). When the diver reaches the water level, the cart's x-coordinate is

$$-240 + 15(W + F)$$

where $F = \sqrt{\frac{57 + 50 \sin 9W}{16}}$.

- ## *Part II: Periodic Problems*

You can give overheads to a couple of groups and have them choose one or two of their most interesting examples to share with the class. They should present the situation, give the period, and show a sketch of the graph.

Here are some of the many ideas they might mention.

- Phases of the moon

- Menstrual cycles

- The movement of the hands of a clock

- The height of the sun in the sky

2. *Generalizing the Platform*

Tell students that the next activity, *Generalizing the Platform*, is basically a cosine version of earlier work with the sine function. Our approach to extending the cosine function beyond the right-triangle definition is to have students seek to define the cosine function for non-acute angles so that the equation $x = 50 \cos 9t$ will give the platform's horizontal position for all values of t.

3. Discussion of *Generalizing the Platform*

Have a club card student present Question 1. Discussion of this problem should give you a good sense of how well students understand the process they used for generalizing the sine function (see Days 4 and 5). They should see that the platform's x-coordinate is $-50 \cos 72°$ for $t = 12$. For the equation $x = 50 \cos 9t$ to give this value when $t = 12$, they need to define $\cos 108°$ to be equal to $-\cos 72°$.

Use your judgment about whether you need to discuss Question 2 as well.

• *Question 3*

Let another club card student present Question 3, and provide additional hints and review as needed. The goal is to have students see that in a diagram such as the one shown here, with θ a first-quadrant angle, $\cos \theta$ is equal to $\frac{x}{r}$, and that using this ratio as the general definition for arbitrary angles is consistent with their work in Questions 1 and 2. (If needed, you can go through details such as those used in *Extending the Sine* for the sine function—see Day 4.)

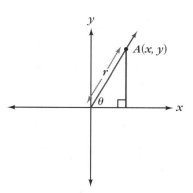

4. Defining the Cosine Function

When students seem clear about the process, you can simply tell them that this is the basis for the formal definition of the cosine function for arbitrary angles. You should post this together with an appropriate diagram (such as the one on the next page).

For any angle θ, we define $\cos \theta$ by first drawing the ray that makes a counterclockwise angle θ with the positive x-axis and choosing a point A on this ray (other than the origin) with coordinates (x, y).

Using the shorthand $r = \sqrt{x^2 + y^2}$, we then define the cosine function by the equation

$$\cos \theta = \frac{x}{r}$$

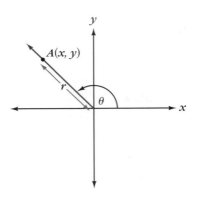

As with the discussion of the sine function, emphasize how well this works, particularly in terms of sign: the extended cosine function is positive when x is positive and negative when x is negative. These two cases correspond exactly to the sign of the platform's x-coordinate, which is positive to the right of center and negative to the left of center.

- ### *The cosine function is well defined*

    Briefly point out that the ratio $\frac{x}{r}$ does not depend on the specific point chosen on the defining ray. You can discuss the fact that this can be proved using similarity, just as was done for the sine function.

- ### *Cosine graphs*

    Have students graph the function defined by the expression $50 \cos 9t$ on their calculators, adjusting the viewing window to include all values from $t = 0$ to $t = 80$. Have them compare the result to their graphs from Question 2 of *Homework 11: Where Does He Land?* They should see that the graphs are the same.

    Also have students graph the "plain" cosine function on their calculators, including negative values for the angle, and compare it with the graph of the "plain" sine function. They should see that the two graphs are identical in shape, but one is "shifted" from the other. You can post and label a graph like the one shown here near the graph of the sine function for future reference.

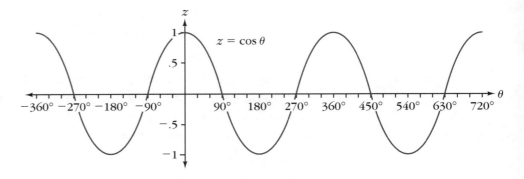

5. Cosine and the Ferris Wheel

Ask students to summarize what this general definition of the cosine function means in terms of the x-coordinate of the platform as the Ferris wheel turns. They should be able to articulate a conclusion like this.

If a Ferris wheel of radius 50 feet makes a complete turn every 40 seconds, starting from the 3 o'clock position, then the x-coordinate of the platform, after t seconds, is given by the function

$$x = 50 \cos 9t$$

Post this along with the previous result on the platform's height.

- *The diver's position when he reaches the water level*

Ask how this conclusion fits into the solution of the unit problem. Bring out that (at least for the current simplified version of the problem), the diver's x-coordinate as he falls is the same as the x-coordinate of the platform when the diver is dropped. In other words:

If the diver is dropped after the Ferris wheel has been turning for W seconds, starting from the 3 o'clock position, then his x-coordinate as he falls is given by the function

$$x = 50 \cos 9W$$

Homework 13: Planning for Formulas

This assignment asks students to sum up and explain their work thus far in the unit.

CLASSWORK

Generalizing the Platform

If the Ferris wheel platform starts at the 3 o'clock position, with the Ferris wheel turning counterclockwise at a constant angular speed of 9 degrees per second, then the platform will remain in the first quadrant through $t = 10$.

During this time interval, the platform's x-coordinate is given by the formula

$$x = 50 \cos 9t$$

This formula specifically uses the fact that the radius of the Ferris wheel is 50 feet and that the angular speed is 9 degrees per second. But the right-triangle definition of the cosine function applies only to acute angles, so this formula isn't defined if t is greater than 10. Your task in this activity is to explore how to extend the definition of the cosine function.

Specific Cases

1. Consider the case $t = 12$.

    a. Find the platform's x-coordinate when $t = 12$. (This was Question 1c of *Homework 11: Where Does He Land?*) You may want to express your answer in terms of the cosine of some acute angle.

    b. What value should you assign to $\cos (9 \cdot 12)$ so that the formula $x = 50 \cos 9t$ gives your answer from Question 1a when you substitute 12 for t?

Continued on next page

2. Consider the case $t = 26$.

    a. Find the platform's x-coordinate when $t = 26$. (This was Question 1d of *Homework 11: Where Does He Land?*) You may want to express your answer in terms of the cosine of some acute angle.

    b. What value should you assign to $\cos(9 \cdot 26)$ so that the formula $x = 50 \cos 9t$ gives your answer from Question 2a when you substitute 26 for t?

The General Case

3. How can you define $\cos \theta$ in a way that makes sense for all angles and that gives the results you needed in Questions 1b and 2b? (You may want to look at *Extending the Sine*.)

Planning for Formulas

You now have all the parts to the puzzle. You simply have to put the pieces together.

You have formulas that tell you each of these things.

- The diver's height at the moment when he is released

- The diver's *x*-coordinate at the moment when he is released

- The amount of time the diver is falling

- The cart's *x*-coordinate when the diver reaches the water level

All of these quantities can be expressed in terms of *W*, which represents the amount of time the Ferris wheel has been turning at the moment when the diver is released.

Write out each of the formulas just described, and explain each of them clearly, including how the general definitions of sine and cosine and principles about falling objects are used in your formulas.

Continued on next page

Also discuss how each of these facts fits into your formulas.

• The Ferris wheel has a radius of 50 feet.

• The center of the Ferris wheel is 65 feet above the ground.

• The Ferris wheel turns counterclockwise at a constant rate, making a complete turn every 40 seconds.

• When the cart starts moving, it is 240 feet to the left of the base of the Ferris wheel.

• The cart moves to the right along the track at a constant speed of 15 feet per second.

• The water level in the cart is 8 feet above the ground.

• When the cart starts moving, the platform is at the 3 o'clock position.

DAYS
14-16

Finding the Release Time

This page in the student book introduces Days 14 through 16.

Shaleen Nand ponders how to pull together all the information she's developed and the different formulas she's collected.

You have developed a large collection of formulas that explain specific parts of the Ferris wheel problem. Now the time has come to put them all together.

Moving Cart, Turning Ferris Wheel

Mathematical Topics

- Solving the simplified version of the unit problem

Students begin to solve the simplified version of the unit problem.

Outline of the Day

In Class

1. Discuss *Homework 13: Planning for Formulas*

- Discuss the individual formulas but not how they fit together

2. *Moving Cart, Turning Ferris Wheel*

- Students solve the simplified version of the unit problem
- Work on this activity continues on Day 15
- The activity is discussed on Day 16

At Home

Homework 14: Putting the Cart Before the Ferris Wheel

1. Discussion of *Homework 13: Planning for Formulas*

Have several students present elements of each formula. You might have one student give the formula itself and have others explain how the facts about the cart and Ferris wheel, the definitions of the sine and cosine functions, and the principles about falling objects fit in. This discussion involves a thorough review of the unit so far, and so it is likely to take much of the class period.

2. Moving Cart, Turning Ferris Wheel

The next activity, *Moving Cart, Turning Ferris Wheel*, is the culmination of the simplified version of the unit problem. No further introduction is needed. Groups will have the rest of today and all of tomorrow to work on this; it will be discussed on Day 16.

If students seem stuck, suggest that they take a guess for *W* and see what happens to both the diver and the cart if the diver is released after *W* seconds.

Homework 14: Putting the Cart Before the Ferris Wheel

In this assignment, students look at a question that is similar to the central unit problem but much simpler.

Working on this simpler problem may help some students with *Moving Cart, Turning Ferris Wheel*.

"I have never seen students so focused on a problem as when we were working on Moving Cart, Turning Ferris Wheel, *the simplified version of the unit problem. One student got on the floor, face down, and pounded it with his fist when he got really close but wasn't satisfied with his result. Another, upon reaching a solution he knew was correct, jumped up and down and screamed with satisfaction. Shortly afterward, a girl commented, 'Boy, if someone from outside walked in, they would think we are really strange for getting so emotional about our math.' The energy in that room was incredible and, at the end of the task, students had an evident sense of accomplishment."*

IMP teacher Susan Ford

Moving Cart, Turning Ferris Wheel

Your job is to figure out when the assistant should let go of the diver. Let $t = 0$ represent the moment when the platform passes the 3 o'clock position. Let W represent the number of seconds until the release of the diver. You need to determine the right value for W.

In addition to giving the value of W, you should also determine these things.

• Where the platform will be in the Ferris wheel's cycle when the diver is dropped

• Where the cart will be when the diver hits the water

Putting the Cart Before the Ferris Wheel

What if you could change where the cart started? That might make things a little easier.

In this assignment, assume that all the facts about the Ferris wheel and the cart are the same as usual except for the cart's initial position.

Suppose the diver is released exactly 25 seconds after the Ferris wheel began turning from its 3 o'clock position.

1. What is his x-coordinate as he falls?

2. Where should the cart start out so that the diver will fall into the tub of water on the cart? (Assume that the cart still starts to the left of the Ferris wheel and travels to the right at 15 feet per second.)

Continuing to Turn the Ferris Wheel

Students continue work on the simplified version of the unit problem.

Mathematical Topics

- Solving the simplified version of the unit problem

Outline of the Day

In Class

1. Discuss *Homework 14: Putting the Cart Before the Ferris Wheel*

2. Continue work on *Moving Cart, Turning Ferris Wheel* (from Day 14)

- Students should develop an equation in terms of the release time and then begin to solve the equation

At Home

Homework 15: What's Your Cosine?

1. Discussion of *Homework 14: Putting the Cart Before the Ferris Wheel*

You can let volunteers explain each of Questions 1 and 2. Question 1 should be a review of familiar ideas. The diver's *x*-coordinate as he falls (and when he lands) is 50 cos (9 · 25), or approximately −35.4.

- *Question 2*

Students need to put several ideas together to answer Question 2. First, they need to find out how long the diver is in the air. If he is released after 25 seconds on the Ferris wheel, his height off the ground (in feet) will be 65 + 50 sin (9 · 25), or approximately 29.6 feet, and so he will fall 21.6 feet to reach the water level. This means that his falling time (in seconds) will be $\sqrt{\frac{21.6}{16}}$, or approximately 1.16 seconds.

Once they have found this falling time, students can look at the cart. The cart will travel a total of approximately 26.16 seconds (the initial 25 seconds plus 1.16 seconds for the diver's falling time). At 15 feet per second, this means the cart will travel approximately 392.4 feet. Therefore, to be in the right place when the diver reaches the water level, the cart must start 392.4 feet to the left of -35.4, which means the cart must start 427.8 feet to the left of center.

2. Continued Work on *Moving Cart, Turning Ferris Wheel*

Have students continue work on *Moving Cart, Turning Ferris Wheel.* Although some groups may have started by guessing specific numbers, you should help them move toward the formulation of a specific equation that they can use to find *W*.

> ● *Hints as students work*
>
> As students work on this, you can remind them to focus on the idea of having the cart and the diver be in the same place when the diver reaches the water level. You can also suggest that they look around the room (or in their notes, or at their work on *Homework 13: Planning for Formulas*) for summaries of key elements of the analysis.
>
> There are several forms for an equation that states that the cart is in the right position when the diver reaches the water level. This rather formidable equation is one possibility:
>
> $$-240 + 15\left(W + \sqrt{\frac{57 + 50\sin 9W}{16}}\right) = 50\cos 9W$$
>
> Once students get this or an equivalent equation, they might try to solve it algebraically. But they will probably find quickly that they are getting nowhere. Tomorrow's discussion suggests several approaches for getting a numerical estimate of the solution.

Homework 15: What's Your Cosine?

> This assignment will give students experience with the new extended definition of the cosine function.

What's Your Cosine?

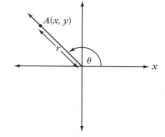

You have seen that we define the cosine function in a manner similar to that for the sine function. If θ is any angle, we draw a ray from the origin, making a counterclockwise angle of that size with the positive x-axis, pick a point (x, y) on the ray (other than the origin), and define r as the distance from (x, y) to the origin, so $r = \sqrt{x^2 + y^2}$. We then define the cosine function for all angles by the equation

$$\cos \theta = \frac{x}{r}$$

As with the sine function, this definition gives the same values for acute angles as the right-triangle definition. Also like the sine function, the extended cosine function can give the same value for different angles.

1. Draw the graph of the function defined by the equation $z = \cos \theta$ for values of θ from $-360°$ to $720°$, and answer these questions.

    a. What is the amplitude of this function?

    b. What is the period of this function? Why is the cosine function periodic?

    c. What are the θ-intercepts of the graph?

    d. What values of θ make $\cos \theta$ a maximum? What values of θ make $\cos \theta$ a minimum?

Continued on next page

2. The questions here are similar to questions about the sine function in *Homework 7: More Beach Adventures*. As in that assignment, your solutions should all be between $-360°$ and $360°$. In Questions 2a and 2b, find exact values for θ. In Questions 2c and 2d, give θ to the nearest degree.

a. Find three values of θ, other than $81°$, such that $\cos \theta = \cos 81°$.

b. Find three values of θ such that $\cos \theta = -\cos 20°$.

c. Find three values of θ such that $\cos \theta = 0.3$.

d. Find three values of θ such that $\cos \theta = -0.48$.

Moving Cart, Turning Ferris Wheel Concluded

Students complete work on the simplified version of the unit problem.

Mathematical Topics

• Completing the simplified version of the unit problem

Outline of the Day

In Class

1. Discuss *Homework 15: What's Your Cosine?*

  • Discuss the sign of the cosine function in each quadrant

2. Discuss *Moving Cart, Turning Ferris Wheel* (begun on Day 14)

  • Go over how the various components of the analysis fit together to create an equation for solving the problem

  • Discuss ways to solve the equation

3. Tell students that they now need to examine a more accurate and more complex version of the unit problem

At Home

Homework 16: Find the Ferris Wheel

---

1. Discussion of *Homework 15: What's Your Cosine?*

You can let volunteers each answer a component of the assignment. The discussion of Question 1 can be similar to that for the Day 6 activity, *The "Plain" Sine Graph.*

• *Question 2*

Use the discussion of Question 2 to get students to generalize about the sign of the cosine function. By looking at the sign of the *x*-coordinate, they should be able to determine that the cosine function comes out positive if the angle is in

---

the first or fourth quadrant, and negative if the angle is in the second or third quadrant. You may find it helpful to use a diagram like the one shown here.

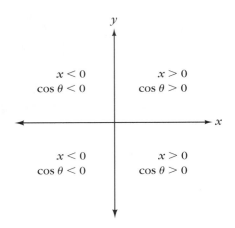

2. Discussion of *Moving Cart, Turning Ferris Wheel*

Comment: The discussion of *Moving Cart, Turning Ferris Wheel* is the culmination of a long effort to solve one version of the unit problem.

It is important that you allow students sufficient time to work through this problem to their satisfaction. Although this activity is "nothing more than" a combination of previous ideas, the process of bringing these varied ideas together may be difficult for many students.

Let spade card students present their analyses of the problem. Give further hints as needed to get an equation that expresses the fact that the cart is in the right place. As noted yesterday, here is one possibility.

$$-240 + 15\left(W + \sqrt{\frac{57 + 50 \sin 9W}{16}}\right) = 50 \cos 9W$$

Make sure students can articulate that the left side of the equation gives the x-coordinate of the cart at the time the diver is at water level, and the right side of the equation gives the x-coordinate of the diver at that same time. In other words, the equation is saying that the diver is landing in the water.

Students may come up with variations on this equation. Let students present any alternate approaches on the overhead, and ask them to articulate their meaning.

Help students appreciate that this analysis is "quadrant-free." That is, it works no matter where the platform is when the diver is released. You might remind them that in the initial activities of the unit (such as *Homework 2: As the Ferris Wheel Turns*), they did not have the general sine and cosine functions, and their analysis was slightly different for each quadrant.

- *Solving the equation*

Here are three ways students might solve the complicated equation just developed.

- Guess-and-check: That is, pick a value for *W*, evaluate both sides, and then repeatedly adjust *W* to bring the two sides of the equation closer together.

- Graphing: For instance, graph the two functions defined by the expressions on the two sides of the equation and then look at where the graphs meet. This will require adjusting the window settings in order to locate the point of intersection.

- Using a "solve" feature on a calculator.

It turns out that the assistant should release the diver about 12.28 seconds after the cart starts moving.

- *Ta-da!*

Give a cheer! The problem is solved! If students haven't already done so, you will probably want to have them substitute *W* = 12.28 into both sides of the equation to confirm that this value really is correct.

Also, be sure to answer the question posed in *Moving Cart, Turning Ferris Wheel* of where the diver is on the Ferris wheel when the assistant releases him. After 12.28 seconds, the Ferris wheel will have turned 9 · 12.28 ≈ 110.5°, which will place the platform between the 11 o'clock and 12 o'clock positions.

The diver's height off the ground when released is given by the expression 65 + 50 sin 9*W*, which gives a value of about 112 feet. The diver's *x*-coordinate is given by the expression 50 cos 9*W*, which comes out to about −17.5, which means he is about 17.5 feet to the left of center.

Students will probably want to work out some more of the stages in the process for *W* = 12.28. For example, the diver must fall about 112 − 8 = 104 feet, which will take about $\sqrt{\frac{104}{16}} \approx 2.55$ seconds. (That is, *F* = 2.55.) Thus, the cart must travel a total of about 12.28 + 2.55 = 14.83 seconds.

- *How much accuracy is needed?*

"How accurate does the assistant need to be?"

Because the value 12.28 is an approximate solution to a practical problem, you should raise the question of how accurate the assistant needs to be. For instance, if he drops the diver after 12.3 seconds, what will happen?

Students can explore this simply by substituting other values for *W* into the two sides of the equation to see how much effect changes in *W* have on the *x*-coordinates of the cart and the diver. For instance, if *W* = 12.3 seconds, then at the moment the diver reaches the water level, the cart's *x*-coordinate is approximately −17.3 and the diver's *x*-coordinate is approximately −17.7. As long as the tub is of a reasonable size, these few inches should not matter. On the other hand, for $W = 12.4$, the cart's *x*-coordinate is approximately −15.9 and the diver's *x*-coordinate is approximately −18.4, so this represents a difference of several feet, which might be problematic.

3. But There's More to the Problem!

Students will no doubt realize from looking at their textbook that there is more to the unit. Some may have looked ahead and seen that we are about to add another layer of complexity. Or perhaps the idea of the initial speed of the diver as he comes off the Ferris wheel was discussed back on Day 2.

In any case, congratulate students on a partial job well done, and tell them that they are not finished yet. The next several days provide a brief break from the central unit problem, and then *Homework 20: Initial Motion from the Ferris Wheel* will introduce them to the next phase of the unit.

Homework 16: Find the Ferris Wheel

In this assignment, students examine the connection between the function describing a rider's horizontal position and the parameters of the Ferris wheel. They also look at the effect of changes in the function on its graph.

Thelma Ryusaki explains her group's solution to the unit problem to her colleagues during their summer inservice.

Find the Ferris Wheel

1. Imagine that the equations in Questions 1a and 1b are each thought of as describing the *x*-coordinate of a rider on some Ferris wheel in terms of time, where the rider is at the 3 o'clock position when $t = 0$. (Here, *t* is in seconds and *x* is in feet.)

   Give the radius, period, and angular speed of the Ferris wheel that each expression represents. (Recall that *angular speed* is the rate at which the Ferris wheel turns and in this situation is given in degrees per second.)

   a. $x = 25 \cos 10t$

   b. $x = 100 \cos 3t$

2. **a.** Write an expression that would give the *x*-coordinate of a rider on a Ferris wheel that has a smaller radius than the Ferris wheel in Question 1a but a greater angular speed.

   b. Describe how the graph for the expression in Question 2a would differ from the graph in Question 1a.

DAYS 17-20

A Trigonometric Interlude

This page in the student book introduces Days 17 through 20.

The Ferris wheel problem gives Vick Chandra a meaningful context in which to develop trigonometric identities.

Congratulate yourself on a major achievement—finding out when the assistant should release the diver. Unfortunately, the work in the unit thus far has involved a significant simplification of the problem. If the assistant uses the solution from *Moving Cart, Turning Ferris Wheel*, it could cost the diver his life. So there's still quite a bit of work to do on the Ferris wheel situation.

Before turning to that more complex version of the problem, the unit digresses with some further study of trigonometry, including the introduction of polar coordinates and the development of some important general principles, called *identities*.

DAY 17

Polar Coordinates

Mathematical Topics

- Introducing polar coordinates
- Expressing rectangular coordinates in terms of polar coordinates

---

Outline of the Day

In Class

1. Discuss *Homework 16: Find the Ferris Wheel*

2. Introduce polar coordinates
- Bring out that a point has many representations in polar coordinates

3. *Some Polar Practice*
- Students find rectangular coordinates from polar coordinates, and vice versa

4. Discussion of *Some Polar Practice*
- Develop general equations for expressing rectangular coordinates in terms of polar coordinates

- Emphasize that a single point can be represented by polar coordinates in more than one way
- Introduce the use of negative angles, angles greater than 360°, and negative values for *r* in polar coordinates

5. Refer students to *A Polar Summary*

At Home

Homework 17: Polar Coordinates on the Ferris Wheel

---

1. Discussion of *Homework 16: Find the Ferris Wheel*

Questions 1 and 2 offer a chance to confirm that students grasp the connections between the parameters of the Ferris wheel problem and the coefficients in the formula for the *x*-coordinate of the rider on the Ferris wheel.

There may be some confusion over period versus angular speed. The latter is the number that appears in the formula. For example, in Question 1a, the coefficient 10 in the expression $25 \cos 10t$ means that the Ferris wheel turns 10 degrees per second. To get the period, students need to divide 360° by this coefficient (or do some equivalent arithmetic process).

- **Question 2**

    Students will probably all have different answers for Question 2, and you can use two or three to illustrate the options. Be sure to get at least a verbal description of how the graph for such a function would differ from that in Question 1a.

    After discussion of Question 2, use your judgment about whether to take the time to have students actually graph examples involving the variations described. If students seem able to articulate that a smaller radius will lead to a smaller amplitude of the graph and that a faster angular speed will lead to a more "scrunched up" graph, then they probably have a sufficient understanding of the ideas.

2. Polar Coordinates

"How have you been describing the platform's position on the Ferris wheel?"

"Where is the origin of the coordinate system?"

"What information have you been using to get these coordinates?"

Ask students to summarize how they have been describing the platform's (or diver's) position on the Ferris wheel. Emphasize that they have described this position in terms of the platform's height off the ground and its horizontal position relative to the center of the Ferris wheel.

Help students to see that in using height and position as they have done, they have been essentially working with an *xy*-coordinate system whose origin is at ground level, directly below the center of the Ferris wheel.

Then ask students what information they have been using to get those coordinates. Bring out that both coordinates are expressed in terms of the radius of the Ferris wheel and the angle through which the platform has turned. Tell them that because the turning occurs at the center of the Ferris wheel, rather than at ground level, it makes more sense to treat the center of the Ferris wheel itself as the origin.

Next, inform students that in the coordinate plane, the measurements corresponding to the radius of the Ferris wheel and the angle of turn are called the **polar coordinates** of a point and are usually represented by the variables r and θ. (We suggest that you wait until after the next activity, *Some Polar Practice,* before introducing the fact that points in the plane have many polar representations. See the subsection "Multiple answers" later today.)

Illustrate the idea of polar coordinates with a diagram as shown here, clarifying the role of each variable.

- The polar coordinate *r* represents the distance from point *A* to the origin.

- The polar coordinate *θ* represents the counterclockwise angle made between the positive direction of the *x*-axis and the ray from the origin through point *A*.

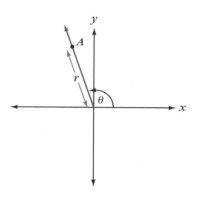

"What are the polar coordinates for a rider on the 50-foot Ferris wheel at the 11 o'clock position?"

Tell students that in giving the polar coordinates of a point, we give the *r* value first, by convention, and illustrate with an example. For instance, ask for the polar coordinates of a rider at the 11 o'clock position on the 50-foot Ferris wheel. Students should see that this position would be represented as (50, 120°).

"What's another name for the system of x- and y-coordinates?"

Point out that with two systems of coordinates under discussion, it's important to be clear which system is being referred to. Review the term *rectangular coordinates* as another name for the system of *x*- and *y*-coordinates. (This term was introduced in the Year 1 unit *Overland Trail*, as was its synonym *Cartesian coordinates*.)

3. *Some Polar Practice*

With the preceding introduction of the basic definitions, have groups work on *Some Polar Practice*.

4. Discussion of *Some Polar Practice*

You can let heart card students present individual problems. Have students discuss whether the answers seem reasonable. For instance, on Question 1a, students should see that both the *x*-coordinate and the *y*-coordinate should be positive (because the point is in the first quadrant) and that the *x*-coordinate should be larger than the *y*-coordinate (because the angle is less than 45°). On Question 1b, they should see that the *x*-coordinate is negative and the *y*-coordinate is positive because the point is in the second quadrant.

To the nearest hundredth, here are the answers.

- For Question 1a: (1.73, 1)
- For Question 1b: (−3.83, 3.21)

The general equations

"In general, how can you find the rectangular coordinates of a point from its polar coordinates?"

Ask students to generalize these examples to develop formulas for finding the rectangular coordinates of a point in terms of its polar coordinates.

If students need a hint, you might suggest that they look at the general definitions of the sine and cosine functions. Students should see that they can simply multiply each defining equation by r. That is, $\cos \theta = \frac{x}{r}$ (by definition), so $x = r \cos \theta$. Similarly, $y = r \sin \theta$. Point out that these relationships work in all quadrants, and post these equations.

If a point has polar coordinates r and θ, then its rectangular coordinates x and y can be found by the equations

$$x = r \cos \theta$$

and

$$y = r \sin \theta$$

Question 2

For Question 2, students are likely to give only the "obvious" answers for the two examples, which are (approximately) (8.25, 14°) for Question 2a and (9.85, 294°) for Question 2b. (If they give other answers, that will lead smoothly into the next subsection.)

Multiple answers

Tell students that by convention, we allow *any* angle, not merely those from 0° to 360°, to be considered as the angle in polar coordinates. Bring out that an angle of more than 360° simply represents more than one complete turn around the origin. Also clarify that negative angles are measured counterclockwise (from the positive direction of the x-axis).

Have students provide at least a couple of alternate solutions for each of Questions 2a and 2b, one with an angle of more than 360° and one with a negative angle. For instance, the answer to Question 2a could be (8.25, 374°) or (8.25, −346°), and the answer to Question 2b could be (9.85, −66°) or (9.85, 654°). Tell students that because a given point has more than one representation, we sometimes speak of "a polar representation" of a point rather than "the polar coordinates" of the point.

You can use the Ferris wheel model to illustrate this idea of multiple representations, pointing out that a rider will pass the same place many times, which gives many different polar coordinate representations of that position. Students will probably see that these representations all have the same r-coordinate. But they should realize that there are infinitely many values for θ that correspond to a given point.

Tell students that we also allow negative values for r, by going in the opposite direction from the ray defined by θ. Either give an example or elicit one from the class. For instance, help students see that the point in Question 2a could also be represented in polar coordinates by the pair $(-8.25, 194°)$.

Without getting into details, we suggest that you point out that because each point has many polar coordinate representations, there are no simple formulas for getting the polar coordinates of a point from its rectangular coordinates.

5. For Reference: *A Polar Summary*

Refer students to the summary of polar coordinates entitled *A Polar Summary*. You may want to go over these ideas briefly again.

Homework 17: Polar Coordinates on the Ferris Wheel

In this assignment, students look at the relationship between polar coordinates and the Ferris wheel problem.

Some Polar Practice

Polar coordinates and rectangular coordinates provide two ways to describe points in the plane. The questions in this activity focus on the relationships between the two systems.

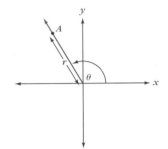

1. **a.** Find the rectangular coordinates for the point whose polar coordinates are $(2, 30°)$.

   b. Find the rectangular coordinates for the point whose polar coordinates are $(5, 140°)$.

2. **a.** Find a pair of polar coordinates for the point whose rectangular coordinates are $(8, 2)$.

   b. Find a pair of polar coordinates for the point whose rectangular coordinates are $(4, -9)$.

A Polar Summary

You know that the position of a point in the plane is usually described in terms of coordinates x and y, which are called its **rectangular coordinates** (or *Cartesian coordinates*). But a point's position in the plane can also be described in terms of **polar coordinates,** usually represented by the letters r and θ.

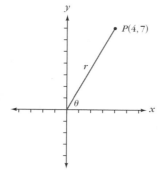

For example, in the accompanying diagram, the point P is shown with rectangular coordinates $(4, 7)$. The variable r represents the distance from P to the origin, and the variable θ represents the angle made between the positive direction of the x-axis and the ray from the origin through P (measured counterclockwise).

You can use the Pythagorean theorem to see that $r = \sqrt{4^2 + 7^2} = \sqrt{65} \approx 8.06$, and you can use one of the trigonometric functions to get θ.

For example, you can use the sine function, whose general definition is $\sin \theta = \frac{y}{r}$. For the point P, this equation becomes $\sin \theta = \frac{7}{\sqrt{65}} \approx 0.868$, which gives $\theta \approx 60°$. In other words, the point P can be represented approximately in polar coordinates as $(8.06, 60°)$.

The process can be reversed, starting from the polar coordinates of a point and finding its rectangular coordinates. For instance, in the next diagram, the point Q is shown with polar coordinates $(10, 240°)$.

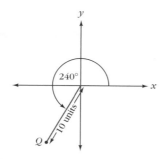

You can use the general definitions of the sine and cosine functions to find the rectangular coordinates of Q.

For example, $\sin \theta = \frac{y}{r}$, so $y = r \sin \theta$. Therefore, the y-coordinate of Q is $10 \sin 240°$, or approximately -8.7.

Continued on next page

Similarly, $\cos \theta = \frac{x}{r}$, which gives $x = r \cos \theta$. Therefore, the x-coordinate of Q is $10 \cos 240°$, which equals -5. Thus, the rectangular coordinates of Q are approximately $(-5, -8.7)$.

Angles Greater Than 360°

The concept of polar coordinates is complicated by the fact that we do not restrict θ to angles between $0°$ and $360°$. An angle of $360°$ or more is simply interpreted as representing more than a complete rotation around the origin. For instance, point P could also be represented in polar coordinates as $(8.06, 420°)$, because a counterclockwise rotation of $420°$ from the positive x-axis leads to the same ray from the origin as a rotation of $60°$. Similarly, point Q could be represented as $(10, 600°)$, $(10, 960°)$, $(10, 1320°)$, and so on.

Negative Angles

We also allow the polar coordinate θ to be negative, by interpreting a negative angle as a *clockwise* rotation from the positive direction of the x-axis. For example, the point Q whose polar coordinates are given as $(10, 240°)$ could also be described by the polar coordinates $(10, -120°)$. The negative sign for the angle of $120°$ means that we are going $120°$ in the *clockwise* direction.

Continued on next page

Negative Values for *r*

The final complication for polar coordinates is that we allow negative values for *r*. If *r* is negative, the point lies in the opposite direction from the point with the corresponding positive *r*-value.

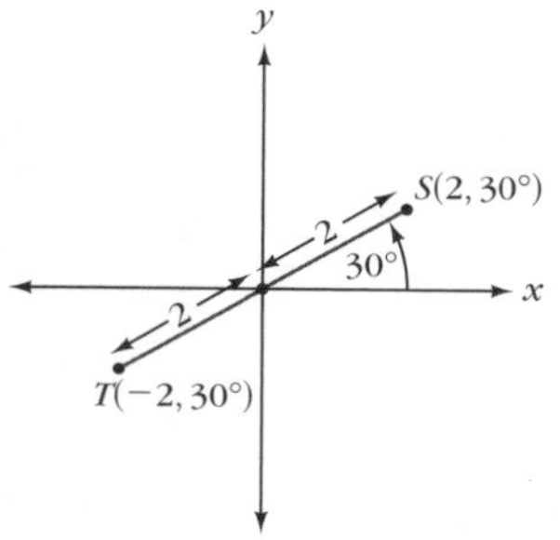

For example, consider the diagram shown here. Suppose point *S*, in the first quadrant, has polar coordinates $(2, 30°)$, and suppose point *T*, in the third quadrant, is in the direction from the origin opposite to *S*, but is also 2 units from the origin. Then *T* can be described by the polar coordinates $(-2, 30°)$.

Comment: Point *T* can also be described by the polar coordinates $(2, 210°)$.

Summary: Multiple Representations

The use of arbitrary angles for θ and of both positive and negative values for *r* means that every point in the plane has infinitely many ways to be represented in polar coordinates. This definitely creates some problems in working with polar coordinates, but it also leads to flexibility. Because points have more than one representation, we sometimes speak of "a polar representation" of a point rather than "the polar coordinates" of the point. For convenience, you might refer to the representation with *r* positive and with θ between 0° and 360° as the "standard" polar representation.

Polar Coordinates on the Ferris Wheel

You may find it helpful to think of polar coordinates in terms of the Ferris wheel. To do so, picture the Ferris wheel with its center at the origin of the coordinate system. Then picture a rider on the circumference of the Ferris wheel, starting on the positive part of the x-axis and going counterclockwise.

In this model, the rider's r-coordinate gives the radius of the Ferris wheel and the rider's θ-coordinate gives the angle through which the rider has turned (starting from the 3 o'clock position). For example, a person on a 30-foot Ferris wheel who has gone one-fourth of the way around has polar coordinates $(30, 90°)$.

1. Suppose a Ferris wheel has a radius of 40 feet and a period of 20 seconds, and the rider passes the 3 o'clock position at $t = 0$. Find both the rectangular coordinates and the "standard" polar coordinates for the rider when $t = 3$ (using the center of the Ferris wheel as the origin).

2. **a.** Find a value of t different from 3 seconds for which the rider would be at the same position as in Question 1.

   b. Use your answer to Question 2a to find a different pair of polar coordinates for the position in Question 1.

3. Find general expressions for both the rectangular coordinates and the polar coordinates of a rider's position at time t (using the Ferris wheel from Question 1 with radius 40 feet and period 20 seconds).

The Pythagorean Identity

Mathematical Topics

- Working with polar and rectangular coordinates
- Developing the Pythagorean identity: $\sin^2 \theta + \cos^2 \theta = 1$

Students develop the Pythagorean identity.

Outline of the Day

In Class

1. Discuss *Homework 17: Polar Coordinates on the Ferris Wheel*

2. *Pythagorean Trigonometry*

- Students develop the identity $\sin^2 \theta + \cos^2 \theta = 1$

3. Discuss *Pythagorean Trigonometry*

- Introduce the term **identity** to describe a general relationship such as the equation $\sin^2 \theta + \cos^2 \theta = 1$
- Review the identity $\sin \theta = \cos (90° - \theta)$

At Home

Homework 18: Coordinate Tangents

1. Discussion of *Homework 17: Polar Coordinates on the Ferris Wheel*

This discussion will tell you how well students are making the connection between polar coordinates and the Ferris wheel. For Question 1, they should see that 3 seconds represents a turn of 54° (based on the period of 20 seconds), so the "obvious" polar coordinates are (40, 54°). To get the rectangular coordinates, students can think in terms of either "coordinate formulas" or "Ferris wheel formulas" to see that the rectangular coordinates can be expressed as (40 cos 54°, 40 sin 54°), which is approximately (23.5, 32.4).

Question 2a illustrates the periodicity of the motion, and students might give values such as 23 seconds, 43 seconds, and so on. For Question 2b, they might use angles of 414°, 774°, and so on. For Question 3, the only "work" needed to get the polar coordinates is the computation that each second represents 18° of turn, so the rider's polar coordinates after t seconds are $(40, 18t°)$. The rectangular coordinates are simply a variation on the formulas students found for the rider's position in the main unit problem.

2. Pythagorean Trigonometry

The Pythagorean identity, $\sin^2 \theta + \cos^2 \theta = 1$, is easily the single most important relationship among the trigonometric functions. This activity will help students discover this relationship.

You may want to review the idea that the definitions of sine and cosine are independent of the choice of point along the appropriate ray and that we sometimes use a point on the unit circle for the definitions.

3. Discussion of *Pythagorean Trigonometry*

Let different diamond card students present results for each question. For Question 1, presenters will probably substitute 1 for r in $\sin \theta = \frac{y}{r}$ and $\cos \theta = \frac{x}{r}$, and come up with $y = \sin \theta$ and $x = \cos \theta$. For Question 2, they should get $x^2 + y^2 = 1$. On Question 3, if students replace x and y as intended, they should get the equation $(\cos \theta)^2 + (\sin \theta)^2 = 1$.

- *A notation convention*

Tell students that by convention, the square of $\cos \theta$ is written as $\cos^2 \theta$ and the square of $\sin \theta$ is written as $\sin^2 \theta$ (and similarly for the other trigonometric functions). Thus, we usually write the equation from Question 3 as

$$\cos^2 \theta + \sin^2 \theta = 1$$

This is a rather strange notational convention, but it is standard, and you may want to check occasionally to verify that students understand that $\cos^2 \theta$ really means $(\cos \theta)^2$, and so on. In particular, if students try to use the "$\cos^2 \theta$" notation on their calculators, they will discover that the calculator will not accept it. For instance, if they want to find $\cos^2 20°$, they will need to enter the expression $(\cos 20)^2$.

• Beyond the unit circle

"Does the relationship $\cos^2\theta + \sin^2\theta = 1$ require you to use points on the unit circle?"

Point out that Questions 1 through 3 are set up using points on the unit circle, and ask whether the relationship $\cos^2\theta + \sin^2\theta = 1$ requires the use of such points. Students might simply point out that the equation doesn't involve r, so once the equation has been established (using the point for which $r = 1$), the value of r no longer matters.

• Question 4

Use your judgment about whether to take class time to go over the verification of the equation for specific values.

• The Pythagorean identity

Tell students that the relationship

$$\cos^2\theta + \sin^2\theta = 1$$

is known as a **Pythagorean identity.** (There are other Pythagorean identities.)

Explain that this equation is called an "identity" because it is true for all angles θ. The "Pythagorean" part of the name comes from its connection with the Pythagorean theorem.

Post this relationship with its name and a statement that this equation is true for all values of θ.

• A familiar trigonometric identity

Students saw the trigonometric identity $\sin\theta = \cos(90° - \theta)$ in the Year 1 unit *Shadows* (see *Homework 24: Your Opposite Is My Adjacent*) and may have reviewed it earlier in this unit. But in previous discussions of this relationship, they knew only the right-triangle definitions of the trigonometric functions. The goal here is to bring out that this identity works for all angles and to give students another illustration of what an "identity" is.

"What other trigonometric identity have you seen involving the sine and the cosine?"

Ask students if they can recall another trigonometric identity involving the sine and the cosine. As a hint, tell them to think of an identity that expresses the sine of an angle in terms of the cosine of a related angle. As a further hint, draw a right triangle and ask how the sine of one base angle might be expressed as the cosine of another angle and what the relationship is between the two angles. Use the fact that the angles are complementary to review the equation $\sin\theta = \cos(90° - \theta)$.

"Does this relationship hold true for all angles?"

Then ask whether this relationship holds true for all angles. You may simply have students verify it for specific values of θ in different quadrants. (Students will examine the generalization of this formula to all quadrants in the next unit, *As the Cube Turns*.)

One possible proof of this relationship uses a quadrant-by-quadrant analysis. A more intuitive approach uses the idea that the graph of the function $y = \cos\theta$ can be obtained by reflecting the graph of the function $y = \sin\theta$ about the line $\theta = 45°$.

Homework 18: Coordinate Tangents

This assignment continues the process of extending the trigonometric functions beyond their right-triangle definitions.

The Ferris wheel gives Jane Kaplan and Erika Cohen a meaningful context in which to develop trigonometric identities.

CLASSWORK

Pythagorean Trigonometry

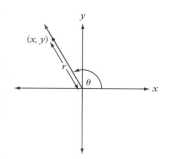

As you have seen, the definitions of the sine and cosine functions are based on a coordinate diagram like the one shown here.

Specifically, to define sin θ and cos θ, we draw a ray from the origin that makes a counterclockwise angle θ with the positive x-axis. Then we pick some point on that ray (other than the origin). We use r to represent the distance from the point to the origin, so $r = \sqrt{x^2 + y^2}$.

If the point has rectangular coordinates (x, y), then we define sin θ as the ratio $\frac{y}{r}$ and define cos θ as the ratio $\frac{x}{r}$.

Because the ratios $\frac{y}{r}$ and $\frac{x}{r}$ are the same no matter which point on the ray is chosen, we can choose any point that is convenient. One common simplification is to pick the point that lies on the unit circle, which is the circle with radius 1 and center at the origin. Choosing this point simplifies matters because it means that $r = 1$.

1. If the point (x, y) is chosen so that it is on the unit circle, how can you express x and y in terms of sin θ and cos θ?

2. What is the equation of the unit circle? That is, what condition must x and y satisfy if (x, y) is 1 unit from the origin?

3. Use your answers to Questions 1 and 2 to write an equation relating sin θ and cos θ for points on the unit circle.

4. Choose four different values of θ, one in each quadrant, and verify on a calculator in each case that your equation in Question 3 holds true.

Coordinate Tangents

You've developed a way to define the sine and cosine functions for arbitrary angles. Now it's time to look at the tangent.

Reminder: For a right triangle such as the one shown here, we define tan θ by the formula

$$\tan \theta = \frac{\text{opposite}}{\text{adjacent}}$$

where "opposite" means the length of $\overline{AC}$ and "adjacent" means the length of $\overline{BC}$.

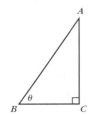

1. Suppose a point in the plane has rectangular coordinates (x, y) and polar coordinates (r, θ), as in the next diagram. How would you define tan θ in terms of x and y? Explain and justify your decision.

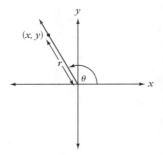

2. It's helpful to have equations connecting the different trigonometric functions. How can you express tan θ in terms of sin θ and cos θ, rather than in terms of the coordinates x and y? (*Hint:* Think about how x and y might be expressed in terms of sine, cosine, and r.)

3. Find each of these values, based on your definition in Question 1.

   a. tan $120°$

   b. tan $230°$

   c. tan $(-50°)$

   d. tan $385°$

4. Sketch a graph of the equation $z = \tan t$, using t for the horizontal axis and z for the vertical axis. Your graph should include values for t from $-180°$ through $360°$.

Positions on the Ferris Wheel

Mathematical Topics

- Extending the definition of the tangent function to all angles
- Developing the identity $\sin \theta = \sin (180° - \theta)$

Outline of the Day

In Class

1. Select presenters for tomorrow's discussion of *POW 2: Paving Patterns*

2. Discuss *Homework 18: Coordinate Tangents*
- Go over the graph of the function

3. *Positions on the Ferris Wheel*
- Students develop the identity $\sin \theta = \sin (180° - \theta)$

4. Discuss *Positions on the Ferris Wheel*

At Home

Homework 19: More Positions on the Ferris Wheel

---

1. POW Presentation Preparation

Presentations of *POW 2: Paving Patterns* are scheduled for tomorrow. Choose three students to make POW presentations, and give them overhead transparencies and pens to take home to use for preparing presentations.

2. Discussion of *Homework 18: Coordinate Tangents*

Let a volunteer present Question 1, but be sure to determine how well students were able to handle this question. Some may have considered it easy after the work with sine and cosine, but others may have been at a loss as to how to proceed. Go over the use of a first-quadrant example to illustrate that the ratio $\frac{y}{x}$ comes from the right-triangle definition and that it makes sense to define $\tan \theta$ in general using this ratio.

Ask whether there are any difficulties that could arise from using this ratio to define the tangent function. If a further hint is needed, ask what happens if $x = 0$. Bring out that the ratio is undefined in that case, and tell students that the tangent function is thus undefined for certain angles.

Have the class determine for which angles the tangent function is undefined. They should see that it is undefined for 90° and 270° in the "first cycle." They might see, more generally, that it is undefined for any odd multiple of 90°. Bring out that for right triangles, the ratio $\frac{\text{opposite}}{\text{adjacent}}$ gets larger and larger as the base angle gets closer to 90°, so it makes sense that there would be a problem at 90°.

On Question 2, the goal is to develop the identity $\tan \theta = \frac{\sin \theta}{\cos \theta}$. Students might get this by writing x and y in terms of r and θ, so that the ratio $\frac{y}{x}$ becomes $\frac{r \sin \theta}{r \cos \theta}$, which then simplifies to $\frac{\sin \theta}{\cos \theta}$.

The examples in Question 3 should be fairly direct applications of the definition. If some students were unable to develop the general definition for the tangent function, you may want to have the class work on Question 3 now. The intent is for students to use right-triangle diagrams and reference angles to find the rectangular coordinates of appropriate points, and to apply the general definition. You can have them verify that their calculators give the same answers.

- ## Question 4

On Question 4, the main goal is to have students see the general pattern of the graph. Take this opportunity to review that the tangent function is undefined for certain angles. Also focus on the sign of the tangent function in the different quadrants. You might make a diagram like the one shown here to go with similar diagrams for sine and cosine.

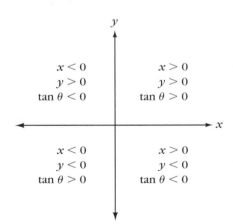

The graph itself should look something like the next diagram.

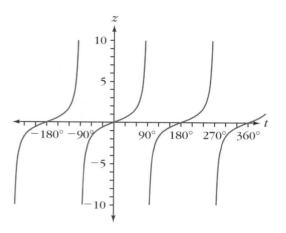

"What is the period of the tangent function?"

Ask what the period of the tangent function is. Students may assume that the period is 360°, as with sine and cosine. Bring out that the tangent function actually has a period of only 180° because of the way the signs work out.

Note: As pointed out in the Year 3 unit *Orchard Hideout,* there is a connection between the use of the word *tangent* in the sense of *tangent to a circle* and the use of the word *tangent* in trigonometry. In the diagram shown here, if the circle has radius 1, then the length of segment $\overline{AB}$, which is part of the line tangent to the circle at A, is equal to $\tan \theta$. If you have not mentioned this connection before, you may want to do so now.

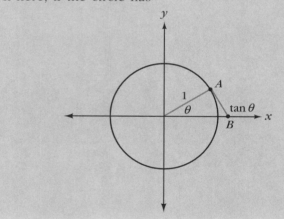

3. Positions on the Ferris Wheel

In this activity, students develop some other trigonometric identities. They will use some of these identities in the next unit, *As the Cube Turns.*

You may want to go over the "reading" portion of this as a whole class, and then have groups begin work on the specific questions.

4. Discussion of *Positions on the Ferris Wheel*

For Question 1, students need to see that the angle for point B is $180° − θ$. Be sure to get an explanation for this conclusion. For instance, students might see that the angle between the ray through B and the negative end of the x-axis must be equal to $θ$, as shown in the next diagram. (They might explain this using the two right triangles in the diagram.)

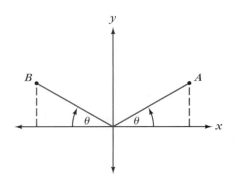

For Question 2, students need only set the two expressions 50 sin $θ$ and 50 sin $(180° − θ)$ equal to each other. Presumably, they will divide by 50 to get the identity

$$\sin θ = \sin (180° − θ)$$

You may want to post this result for reference in later units.

Homework 19: More Positions on the Ferris Wheel

This assignment continues the work of today's activity, *Positions on the Ferris Wheel*.

CLASSWORK

Positions on the Ferris Wheel

In *Pythagorean Trigonometry,* you developed the equation $\cos^2 \theta + \sin^2 \theta = 1$. You saw that this equation is true no matter what value is substituted for the angle θ.

Equations with variables that are true no matter what values are substituted for the variables are called **identities.** In this assignment and in *Homework 19: More Positions on the Ferris Wheel,* you will look at other identities involving the sine and cosine functions. The Ferris wheel model can help in developing and understanding these identities.

The diagram below shows two riders on a Ferris wheel, one at the 1 o'clock position and the other at the 11 o'clock position. The first rider has turned 60° from the 3 o'clock position, and the second has turned 120°.

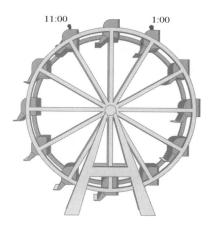

Continued on next page

If the radius of the Ferris wheel is 50 feet, then a rider's height, compared to the center of the Ferris wheel, is given by the expression 50 sin θ. But these two riders are at the same height, so 50 sin 60° = 50 sin 120°. Dividing by 50 gives the relationship

$$\sin 60° = \sin 120°$$

The equation sin 60° = sin 120° can be generalized, using the next diagram, to get an identity involving the sine function.

In this diagram, points A and B represent the positions of two riders on a Ferris wheel, so A and B are the same distance from the origin.

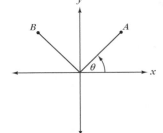

The angle θ represents the angle of turn for a rider at point A. You should assume that A and B are at the same height on the Ferris wheel, so the two points have the same y-coordinate.

1. Find the angle through which the rider at point B has turned. You should express your answer in terms of θ. (*Hint:* Consider an example, such as $\theta = 20°$, find the angle for point B, and then generalize.)

2. Use the fact that A and B are at the same height to write a generalization of the equation sin 60° = sin 120°.

More Positions on the Ferris Wheel

This assignment continues the topic of trigonometric identities.

Part I: Clockwise and Counterclockwise

The diagrams here show two Ferris wheel riders. Both started from the 3 o'clock position, but the first turned 30° and is now at the 2 o'clock position, while the second turned −30° and is now at the 4 o'clock position. (Recall that negative angles are interpreted as clockwise motion.)

Assume that the radius of the Ferris wheel is 50 feet, and recall that if a rider turns through an angle θ, then his x-coordinate is given by the expression $50 \cos \theta$. Thus, the first rider's x-coordinate is $50 \cos 30°$ and the second rider's x-coordinate is $50 \cos (-30°)$.

1. a. Explain why these two riders have the same x-coordinate. That is, why are they the same distance to the right of the center of the Ferris wheel?

    b. What does the result in Question 1a tell you about $\cos 30°$ and $\cos (-30°)$?

Now consider the general situation. In the diagram shown on the next page, points C and D represent two positions on the same Ferris wheel. In the case of point C, the rider has turned through an angle θ. For point D, the rider has turned through an angle $-\theta$.

Continued on next page

2. a. Explain why points C and D have the same x-coordinate.

b. Use the diagram and Question 2a to explain why $\cos \theta$ and $\cos (-\theta)$ must be equal.

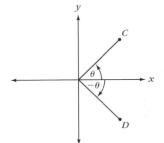

Part II: From Identity to the Ferris Wheel

In Part I, you started with a situation on the Ferris wheel and generalized it to get a trigonometric identity. In this part of the assignment, your task is to start with the identity and create a Ferris wheel explanation.

Consider the equation $\sin (-\theta) = -\sin \theta$.

3. Substitute specific values for θ to confirm that the equation is true for those values. Try a variety of values, including angles that are negative and angles that are greater than 360°.

4. Create a Ferris wheel situation to explain the equation in a manner similar to that used in Part I.

5. Create a coordinate system diagram like that used in Part I to illustrate the situation.

POW 2 Presentations

Mathematical Topics

- Developing the trigonometric identities $\cos(-\theta) = \cos\theta$ and $\sin(-\theta) = -\sin\theta$
- Analyzing a problem involving the Fibonacci sequence
- Using a recursion equation

Outline of the Day

In Class

1. Discuss *Homework 19: More Positions on the Ferris Wheel*

2. Presentations of *POW 2: Paving Patterns*

- Have students explain the recursion equation in terms of the problem situation

At Home

Homework 20: Initial Motion from the Ferris Wheel

1. Discussion of *Homework 19: More Positions on the Ferris Wheel*

You can have different volunteers present their ideas on each of the questions in the assignment.

- *Part I: Clockwise and Counterclockwise*

Part I is similar to yesterday's activity, *Positions on the Ferris Wheel*. For Question 1a, students might use a diagram like the first one on the next page, with *C* and *D* representing the positions of the two riders. They can show that the two right triangles are congruent (or use a more intuitive argument) to explain why *C* and *D* have the same *x*-coordinate.

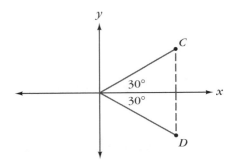

To answer Question 1b, they should reason that the formula gives 50 cos 30° and 50 cos (−30°) as the two *x*-coordinates. Because these *x*-coordinates are equal, cos 30° and cos (−30°) must be equal.

The next diagram simply generalizes the previous one.

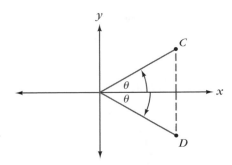

The explanation for Question 2a is essentially the same as that for Question 1a. For Question 2b, students will probably have little difficulty going from their work in deriving the equation cos (−30°) = cos 30° to the more general identity

$$\cos(-\theta) = \cos\theta$$

"How is this equation connected to the graph of the cosine function?"

Ask students how this equation is connected to the graph of the cosine function. Help them see that the equation says that the graph of the function $y = \cos x$ is symmetric about the *y*-axis.

• Part II: From Identity to the Ferris Wheel

For Part II, begin by having students illustrate their work from Question 3, explaining the identity sin (−θ) = −sin θ. Be sure to include an example involving a negative angle, because the issue of signs is important here. Specifically, they should see that if θ itself is negative, then −θ is positive.

The main task of Part II is finding a Ferris wheel situation that expresses the given identity. Let volunteers offer ideas. For example, a student might use the diagram from Part I and point out that a rider going counterclockwise is just as high above the center of the Ferris wheel as a rider going clockwise is below the center.

Post all of the identities discussed in this assignment and save them for use in the next unit, *As the Cube Turns*.

2. Presentations of *POW 2: Paving Patterns*

Let the three students make their presentations, and have other students add their ideas.

If students make a table of results, showing the number of ways to pave an $n \times 2$ path, it will probably look something like this:

Length of path	Number of paving patterns
1 foot	1
2 feet	2
3 feet	3
4 feet	5
5 feet	8
6 feet	13

For instance, here are the eight different ways in which a 5×2 rectangle can be paved:

There are several important aspects to this problem. The first stage is students' ability to organize their lists of paving arrangement so that they get the right data in the table. It may be productive for them to share ways in which they avoided omitting any arrangements.

The second stage is recognizing the pattern in the table. The sequence 1, 2, 3, 5, 8, 13, . . . is a slight variation on the sequence known as the **Fibonacci sequence.** (The Fibonacci sequence begins with two 1's instead of one. That is, it goes 1, 1, 2, 3, 5, 8, and so on.)

Leonardo Fibonacci (c. 1170–1240), also known as Leonardo of Pisa, was a major mathematician of the Middle Ages. He studied with an Arab master while his father served as consul in North Africa. In his first book, *Liber Abaci* (Book of the Abacus), published in 1202, he made the Hindu-Arabic numeral system—the base 10 place value system—generally available in Europe. Prior to that time, the system was known in Europe only to a few intellectuals who had seen translations of the writings of the ninth-century Arab mathematician and astronomer al-Khwarizmi.

Liber Abaci also contained a discussion of the number sequence that now bears Fibonacci's name, introducing the sequence in connection with this problem:

A certain man put a pair of rabbits in a place surrounded on all sides by a wall. How many pairs of rabbits can be produced from that pair in a year if it is supposed that every month each pair begets a new pair which from the second month on becomes productive?

• *The pattern in the table*

As students will probably see, each number in the right-hand column of the table is the sum of the two preceding numbers. For instance, the entry 13 is the sum of the two preceding terms, 8 and 5.

Formally, if we let a_n represent the number of ways to pave an $n \times 2$ rectangle, then the pattern in the table can be represented by the equation

$$a_n = a_{n-1} + a_{n-2}$$

For example, the case $n = 6$ gives us the formula $a_6 = a_5 + a_4$, which is our earlier relationship $13 = 8 + 5$. (Because a_n is defined only for positive values of n, the equation $a_n = a_{n-1} + a_{n-2}$ makes sense only if $n \geq 3$.)

Comment: Students may remember seeing something similar to this in the Year 3 unit *Pennant Fever* when they explored Pascal's triangle, in which each term is the sum of two terms from the previous row.

"What is this type of formula called?"

Ask students if they can recall from *POW 1: The Tower of Hanoi* what this type of formula is called. If necessary, remind them of the term *recursion equation*.

"What does this equation say about the paving patterns?"

Ask students what this recursion equation says in terms of the paving patterns. Bring out that it says that the number of $n \times 2$ paths is the sum of the number of $(n - 1) \times 2$ paths and the number of $(n - 2) \times 2$ paths.

Students can continue the table using the recursion equation. For instance,

$$a_7 = a_6 + a_5 = 13 + 8 = 21$$

$$a_8 = a_7 + a_6 = 21 + 13 = 34$$

$$a_9 = a_8 + a_7 = 34 + 21 = 55$$

They can continue in this way to get the equation

$$a_{20} = a_{19} + a_{18} = 6765 + 4181 = 10{,}946$$

Thus, there are nearly 11,000 ways for Al and Betty to pave a 20×2 path with their 1×2 paving stones.

• Explaining the recursion equation

"Why does this recursion equation hold true?"

An important aspect of the problem is understanding why this particular recursion equation should hold true. Ask the class why the number of $n \times 2$ paths should be the sum of the number of $(n - 1) \times 2$ paths and the number of $(n - 2) \times 2$ paths.

If no one can explain this, here is a sequence of questions you can ask to lead students to understand the pattern.

"What can the 'beginning' of the paving pattern look like?"

First, ask what the "beginning" of the paving pattern can look like, bringing out that there are two options. One option is to place a single stone sideways as shown below.

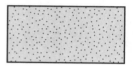

The other option is to place two stones "vertically" adjacent to each other as in the next diagram.

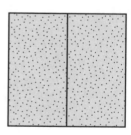

"In each case, how many more feet of path are needed to build a path that is n feet long?"

For each case, ask how many more feet of path are needed to build a path that is n feet long. In the first case, the initial part of the path is only 1 foot long, so an additional $n - 1$ feet are needed. In the second case, the initial part of the path is 2 feet long, so only an additional $n - 2$ feet are needed.

"How many ways are there to complete each possible 'beginning'?"

Then ask how many ways there are to complete the path in each case. Help students to see that the number of $n \times 2$ paths that begin with a horizontal paving stone is a_{n-1}, because the remainder of such a path is simply an $(n - 1) \times 2$ path. Similarly, the number of $n \times 2$ paths that begin with a pair of vertical paving stones at the top is a_{n-2}, because the remainder of the path is simply an $(n - 2) \times 2$ path.

For example, of the eight patterns shown earlier for a 5 × 2 path, there are five with a horizontal paving stone at the top, and three with a pair of vertical paving stones at the top.

Comment: Some students may develop the recursion equation from this general analysis, rather than from the numerical data. There is no right or wrong order in which to think about this.

• *The closed form*

There is a closed-form expression for the number of $n \times 2$ paths, but finding this expression requires knowing (or guessing) that it should be the sum of two exponential expressions. The number of $n \times 2$ paths is given by the expression

$$\frac{1}{\sqrt{5}}\left[\left(\frac{1 + \sqrt{5}}{2}\right)^{n+1} - \left(\frac{1 - \sqrt{5}}{2}\right)^{n+1}\right]$$

Amazingly, this gives a positive integer for every positive integer value of n. To get the nth term of the Fibonacci sequence, simply replace $n + 1$ with n.

Homework 20: Initial Motion from the Ferris Wheel

This short assignment will serve as the introduction to the final phase of the central unit problem.

Initial Motion from the Ferris Wheel

When you finished the activity *Moving Cart, Turning Ferris Wheel*, you may have thought that you had solved the main unit problem. And you were right, sort of. But that activity involved a simplified version of the unit problem, and now you're ready to tackle a more complicated version.

The complication is that when the diver is released from the turning Ferris wheel, he does not fall as if dropped from a stationary Ferris wheel. He is actually moving when he is dropped, and this fact can affect both the speed and

Continued on next page

the path of his fall. The final aspect of the unit problem, then, is to figure out how to take this into account.

To get started, the first problem in this assignment is about a situation that involves circular motion but does not involve gravity.

1. You are a skateboard rider. You go to a park that has a merry-go-round such as the one shown in the picture on the preceding page. You hold onto the railing while the merry-go-round spins rapidly, so you are moving in a circular path. Suddenly, you let go of the railing.

   What is the path of your motion?

   (Assume that your skateboard has perfect ball bearings and that the merry-go-round is surrounded by an ideal surface so that there is no friction anywhere to slow you down.)

2. Now think about the Ferris wheel. You are holding onto a railing on the circumference of a Ferris wheel that is turning rapidly. Suddenly, you let go.

   Sketch the path of your fall for at least one position in each quadrant.

21-25

A Falling Start

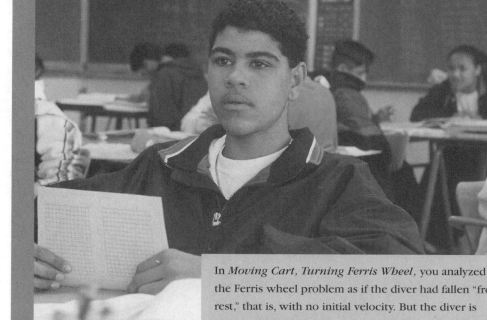

This page in the student book introduces Days 21 through 25.

Mark Martin tries to visualize what effect the motion of the Ferris wheel will have on the diver's path upon release.

In *Moving Cart, Turning Ferris Wheel,* you analyzed the Ferris wheel problem as if the diver had fallen "from rest," that is, with no initial velocity. But the diver is moving while he's still on the Ferris wheel, because the platform itself is moving. This means that the diver will have some initial speed at the moment when he is released.

The simplest cases are those in which the diver's initial motion is vertical—either straight up or straight down. These cases require a new look at the formulas related to falling objects, as well as some new ideas about solving equations.

Leaving Circular Motion

Mathematical Topics

- Seeing that an object released from motion in a circular path continues on a straight path tangent to the circle at the point of release
- Studying the effect of gravity on the path of the diver's fall

Students explore what happens if an object constrained to a circular path is released.

Outline of the Day

In Class

1. Form new random groups
2. Discuss *Homework 20: Initial Motion from the Ferris Wheel*
   - Establish that in the absence of external forces, an object released from a circular path will travel in a straight line
   - Establish that the diver will have an initial speed when released equal to his speed while on the Ferris wheel

3. *Look Out Below!*
   - Students find the falling time of an object that is already moving downward
   - The activity will be discussed on Day 22

At Home

Homework 21: So What Happens to the Diver?

Special Materials Needed

- A transparency of a diagram for use in discussing the motion of an object released from a circular path (see Appendix B)
- (Optional) Materials to demonstrate what happens when an object is released after traveling in a circular path

Discuss With Your Colleagues

What If Their Intuition Is Wrong?

Today's material includes the principle that if an object is being constrained to travel in a circular path and is then released from the constraint, it will continue in a straight path tangent to the circle (assuming there are no other forces acting on the object).

This principle does not fit everyone's intuition, and some students may resist the idea. What do you do about this?

1. Forming New Groups

This is an excellent time to form new random groups. Follow the procedure described in the IMP *Teaching Handbook,* and record the members of each group and the suit for each student.

2. Discussion of *Homework 20: Initial Motion from the Ferris Wheel*

This homework discussion has two main purposes.

- To provide an opportunity for students to share what they *think* will happen in each of the two situations presented

- To let students know what *really* happens in each situation

Unfortunately, the reality does not fit everyone's intuition, and ultimately, you may need simply to insist that students take your word on this. We recommend that you consult your physics department for ideas about experiments or videos that might be helpful here. (One suggestion is included in the optional subsection "An outdoor demonstration.")

- *Question 1*

    Let students share ideas. Many students may think (incorrectly) that the skateboarder will continue in a circular, or at least curving, path, such as shown here. (This diagram and the next one represent a view from above the merry-go-round.)

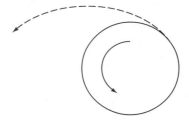

In fact, a person traveling in a circular path and then released would travel in a direction tangent to the circular path, as shown in the next diagram. (This diagram is included in Appendix B.)

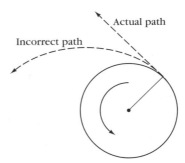

An important aspect of this principle is that the tangent to a circle is perpendicular to the radius of the circle at the point of tangency. (Students saw this geometric fact in the Year 3 unit *Orchard Hideout*.) This fact will be useful in helping students determine certain angles involved in the diver's initial motion as he leaves the Ferris wheel.

As noted previously, you may find some experiments or videos to help convince students of this principle. Or you might use the idea of the game "crack-the-whip" or the discus or hammer throw athletic events to help students visualize the principle, perhaps getting a discus or hammer thrower to demonstrate. Or you could use the experiment suggested in the next subsection.

In any case, you should post the principle, because it will be needed later in the analysis.

In the absence of external forces such as gravity, an object released from a circular path will travel in a straight line. That straight-line path forms a right angle with the circle's radius to the point of release.

• *Optional: An outdoor demonstration*

Note: This demonstration can be done indoors, but for safety considerations, it is probably better to go outside. (You should test it out yourself first before doing a demonstration for the class.) You should explain the experiment before you go outside, and you may want to let students try it themselves.

The demonstration works as follows: Tie an object, such as a roll of masking tape or a chalkboard eraser, to the end of a piece of string (perhaps four feet long). Spin around, swinging the string and object around yourself in a circle (at a more or less constant height). Then let go of the string, and watch it and the object fly. Students need to notice two things about the object's path of flight.

- The object continues in a straight line (except that it sinks to the ground). It does not continue in its original circular path, although many people's intuition suggests that it should.

- That straight line is tangent to the circle. For example, if you were turning counterclockwise (that is, to your left) and facing north when you let go, then the object would fly west.

The object being attached to the string should be heavy enough that there is little air resistance, but soft enough that if someone gets hit, no serious injury will result. Also, you will want to spread out the students as they work so they don't get hit by each other's flying objects. Keep in mind that students may not have good control over the timing of their release of the string.

- ## *An object's initial speed on release*

"How fast would an object be going when first released?"

Ask students *how fast* the object would be going when first released. They may have various ideas on this, and you may have to simply tell them this principle.

> **When an object is released from a circular path, it will have an initial speed equal to the speed at which it was already traveling.**

- ## *Question 2*

The goal here is to get students to build on the discussion from Question 1 and to recognize that the Ferris wheel's motion affects the diver's path. We do not expect students to work out the fine details about that effect. We suggest that you have students describe both what they thought would happen as they worked on the homework and what they now think will happen based on the principle stated for Question 1.

You may find it helpful to have students imagine that the Ferris wheel is turning quite rapidly, so that they can picture the diver being "flung" from the wheel when he is released. That will make this situation similar to the situation in Question 1. As with Question 1, don't be surprised if students have a variety of ideas about what the effect of the motion of the Ferris wheel will be on the path of the diver's fall. It may be helpful to focus discussion on what happens for different release positions.

"What would happen to the diver if he were released at the 12 o'clock position and there were no gravity?"

For example, if the diver were released at the 12 o'clock position, he would be moving to the left at the moment of release. You may want to ask what would happen to him if there were no gravity. Bring out that according to the principle discussed for Question 1, he would continue on a horizontal path, level with the top of the Ferris wheel.

"How is the diver's landing position affected by the motion of the Ferris wheel?"

Acknowledge that gravity *does* affect the diver's path, and ask how his landing position (after being released at the 12 o'clock position) would be different if he were released from a spinning rather than stationary Ferris wheel. Students will probably realize that he would end up farther to the left if released from a spinning Ferris wheel than if he were simply dropped straight down.

"What would happen at the 4 o'clock position without gravity? What about with gravity?"

You can have students do a similar analysis for release at the 4 o'clock position. Try to get agreement that in the absence of gravity, the diver would travel upward to the right (in a direction perpendicular to the radius to the circle at the position of release). Taking gravity into account, students should see that the diver's initial motion will cause him to end up farther to the right than if he were simply dropped. They may also realize that it will take longer for him to reach the ground than if he were simply dropped. The next diagram illustrates what might happen if he were released at about the 4 o'clock position.

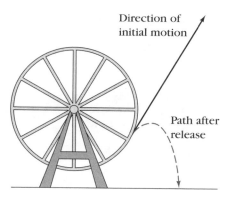

Direction of initial motion

Path after release

The key principle to bring out is that the diver's initial path, like that of the skateboarder, is tangent to the circular path of his motion prior to release. Students may question whether this effect really would happen on a Ferris wheel. As noted previously, if you have them imagine that the Ferris wheel is turning as rapidly as the object in your demonstration, the effect may seem more plausible.

• The diver's speed

"How fast would the diver be going when he is first released?"

Ask about the diver's initial speed when he is released. Students may remember that they worked out earlier in the unit (as Question 1 of *Homework 2: As the Ferris Wheel Turns*) that while the diver is on the Ferris wheel, his speed is 2.5π feet per second, or about 7.85 feet per second. (This speed is a function of the Ferris wheel's angular speed and radius.)

> If you have this result posted from Day 3, students can simply refer to that result. If not, have them reconstruct the analysis.

Combine the result about the diver's speed while on the Ferris wheel with the principle established in Question 1 to get this conclusion.

> **When the diver is released, he will have an initial speed of 2.5π feet per second, which is approximately 7.85 feet per second.**

Post this fact. (You probably have already posted the speed of the turning Ferris wheel, but this statement connects that speed to the diver's speed on release from the Ferris wheel.)

The next two weeks or so are devoted to studying the effect of the turning of the Ferris wheel on the diver's initial path.

The analysis starts with a qualitative look at the situation (*Homework 21: So What Happens to the Diver?*) and then moves on to examine three special cases, each developed through pairs of activities, one involving the Ferris wheel and one set in a different context.

- Initial *downward* motion: *Look Out Below!* (begun today) involves a falling pillow; *Homework 22: Big Push* looks at the Ferris wheel problem for the 9 o'clock position of the diver.

- Initial *upward* motion: *Three O'Clock Drop* (Day 24) looks at the diver's release at the 3 o'clock position; *Homework 24: Up, Down, Splat!* involves a egg container thrown upward.

- Initial *horizontal* motion: *High Noon* (Day 26) concerns the diver's release from the 12 o'clock position; *Homework 26: Leap of Faith* looks at someone jumping straight out off a roof.

Day 23 includes a digression to look at the use of the quadratic formula for solving the quadratic equations that arise in dealing with situations involving falling objects. Day 25 is devoted to completing the work from Day 24.

On Day 27, students begin to work on situations in which the initial velocity is neither purely vertical nor purely horizontal. As preparation, they look at components of velocity in situations that do not involve gravity (*The Ideal Skateboard* and *Homework 27: Racing the River*).

Then they return to the Ferris wheel with a group of activities (*One O'Clock Without Gravity, Homework 28: General Velocities,* and *Moving Diver at Two O'Clock*) that get them ready to complete the unit problem in the culminating activity, *"High Dive" Concluded* (on Day 31).

3. *Look Out Below!*

Look Out Below! applies the ideas of *Free Fall* (Day 9) to a situation in which there is a non-zero initial speed (downward).

• *Review of objects falling from rest*

Before groups begin today's activity, we recommend that you have students review both the idea of constant acceleration and the method of "averaging the endpoints." You may want to do this by looking at the situation in Question 1 of *Free Fall*.

"If an object falls freely from rest, what is its instantaneous speed after 5 seconds? What is its average speed for those 5 seconds? How far does it fall?"

For instance, ask what the instantaneous speed would be after 5 seconds if an object fell freely from rest. Then ask what the object's average speed would be for those 5 seconds. Finally, ask how far the object would fall in those 5 seconds.

The goal here is to review the principle that the object's average speed for the time interval is the average of its initial speed (which is 0, because the object is falling from rest) and its final speed (which is simply 32 · 5 feet per second). Emphasize that this principle holds because the acceleration is constant.

You may also want to review with students that once they have the average speed, they can get the distance traveled by multiplying average speed by time spent traveling.

• Getting started

After this review of the case of falling from rest, have groups look at *Look Out Below!* Point out that although this activity involves another falling object, the rate of acceleration here is different, because air resistance has a significant effect on the motion of the pillow. (Because of this change in acceleration, you may want to go over Question 1 as a whole class, to be sure the situation is clear.)

If students need a hint, have them suppose that the fall took t seconds. They should then find expressions for the pillow's instantaneous speed t seconds after Maxine saw it, for its average speed for the first t seconds, and for the distance it travels in those t seconds.

This hint should lead students to a quadratic expression in t for the distance the pillow falls in t seconds, and they can use a guess-and-check approach to solve the related equation. (This activity will provide a lead-in to discussion of the quadratic formula tomorrow.)

• A disclaimer

In fact, as noted in the activity itself, this problem involves a major oversimplification of the situation, because the air resistance will lead to the pillow not accelerating at a constant rate. After the activity has been discussed, you may want to talk about this oversimplification.

Homework 21: So What Happens to the Diver?

This assignment asks students to apply ideas about release from circular motion to the Ferris wheel situation, at least from an intuitive perspective.

Look Out Below!

It was Thanksgiving vacation, and many students were going home for the holiday. Maxine heard a voice shout, "Hey, up there! Could you toss me my pillow?"

She glanced out the floor-to-ceiling window of her room in the high-rise dormitory, just in time to see the pillow fly past.

Because pillows are comparatively light for their size, the effect of air resistance on a falling pillow cannot be ignored. For this activity, assume that the falling pillow accelerates at a rate of only 20 feet per second for each second it falls (although this is an oversimplification), and assume that the pillow was traveling at an instantaneous speed of 30 feet per second when Maxine saw it.

1. What was the pillow's *instantaneous* speed 1 second after Maxine saw it? Two seconds after Maxine saw it?

2. What was the pillow's *average* speed for the first 2 seconds after Maxine saw it?

3. How far did the pillow fall during the first 2 seconds after Maxine saw it?

Maxine walked over to the window, looked down to the sidewalk, and saw someone reach over to pick up the pillow. The sidewalk was 200 feet below Maxine's window.

4. How long did it take for the pillow to reach the ground from the time Maxine saw it? Give your answer to the nearest tenth of a second.

5. Find a general expression for the height of the pillow *t* seconds after Maxine saw it.

HOMEWORK 21

So What Happens to the Diver?

You have seen something about what happens to objects when they start out moving in a circular path and are then released. In this assignment, you will look at how these principles affect our circus diver.

In thinking about the questions in this assignment, you may want to pay special attention to the cases in which the diver is released from the 3 o'clock, 6 o'clock, 9 o'clock, or 12 o'clock position.

1. From what release positions will the diver's falling time be increased by his initial motion from the Ferris wheel? From what release positions will it be made shorter?

2. From what release positions will the diver land to the left of his release position? From what release positions will he land to the right?

3. In *Moving Cart, Turning Ferris Wheel,* using the assumption that the diver fell straight down as if dropped from rest, you found that the assistant should hold onto the diver for about 12.3 seconds. (This means the diver will be released between the 12 o'clock and 11 o'clock positions on the Ferris wheel.)

If you take into account the effect of the Ferris wheel's motion on the diver's fall, do you think the assistant should hold onto the diver longer than 12.3 seconds or let go sooner than that? Explain your reasoning.

Look Out Below!

Mathematical Topics

- Finding an expression for the position of a falling object with initial downward speed as a function of time

Outline of the Day

In Class

1. Discuss *Homework 21: So What Happens to the Diver?*

2. Discuss *Look Out Below!* (from Day 21)

   - Use "averaging the endpoints" to develop an expression for the pillow's height as a function of time
   - Solve the resulting quadratic equation

At Home

Homework 22: Big Push

1. Discussion of *Homework 21: So What Happens to the Diver?*

On Question 1, bring out that if the diver is moving partly upward at the moment of release, his falling time will be increased by the effect of the motion of the Ferris wheel. But if he is moving partly downward when released, his falling time will be decreased by the effect of the motion of the Ferris wheel.

Students might answer Question 1 either in this sort of descriptive form or more specifically in terms of Ferris wheel positions. For instance, they might say that the diver's falling time is increased if he is on the right side of the Ferris wheel (moving upward from the 6 o'clock position to the 12 o'clock position, counterclockwise).

Similarly, on Question 2, students should see that if the diver is moving to the left when he is released, he will land to the left of his release position, and if he is moving to the right when he is released, he will land to the right of his release position.

Question 3 is considerably more complex, and there is no purely intuitive way to answer it. With the diver being released between the 12 o'clock and 11 o'clock positions on the Ferris wheel, here are the issues.

- On the one hand, the motion of the Ferris wheel will cause the diver to reach the water sooner. This means that the cart will not have traveled as far when the diver reaches the water. This effect, taken by itself, suggests that the assistant should hold onto the diver longer, to allow the cart more time to travel.

- On the other hand, the motion of the Ferris wheel will cause the diver to be farther to the left when he reaches the water, which means the cart doesn't have to go as far. This effect, taken by itself, suggests that the assistant should hold onto the diver for less time.

You may want to let groups discuss this for a few minutes and then share their ideas. Bring out that there is no simple way to see how to balance these two effects. Tell students that they will have to work out the numerical details of the situation in order to answer the question. (Students will actually answer this question in *Homework 30: The Danger of Simplification*.)

2. Discussion of *Look Out Below!*

Let club card students give solutions to the two parts of Question 1. They should be able to come up with the values 50 ft/sec and 70 ft/sec.

On Question 2, the intent is for students to use the "averaging the endpoints" method. Thus, if the pillow traveled for 2 seconds, they should average the beginning speed, which is 30 ft/sec, with the instantaneous speed after 2 seconds, which is 70 ft/sec (from Question 1). This gives an average speed of 50 ft/sec for the interval from 0 to 2 seconds.

For Question 3, get an explicit statement of how to obtain the distance from this average speed, because this is a key step in the generalization required in Question 4. Students need to see that they must multiply the average speed (50 ft/sec, from Question 2) by the length of the time interval (2 seconds) to get that the pillow falls 100 feet during the 2 seconds.

For Question 4, let a spade card student present his or her group's reasoning. If the group used the hint, the presenter will probably begin by explaining that the instantaneous speed after t seconds is $30 + 20t$ ft/sec, so the average speed for the first t seconds is $\frac{30 + (30 + 20t)}{2}$ ft/sec.

We suggest that you have the presenter (or a volunteer) simplify this expression before proceeding. You might have the class verify that the simplified expression, $30 + 10t$, gives the value 50 when $t = 2$, which is consistent with the answer to Question 2.

The next step is to multiply this average speed by the length of the time interval (as discussed for Question 3). This gives the expression $t(30 + 10t)$, or $30t + 10t^2$, as the distance traveled in t seconds. (Again, you might have students check that for $t = 2$, this matches their earlier result.) The presenter will probably then simply set this expression equal to 200 feet (the distance that the pillow needs to fall), to get the equation $30t + 10t^2 = 200$.

"What type of equation is this? What is its standard form?"

We recommend that you have the class identify this as a quadratic equation and put it into standard quadratic form, perhaps simplifying by dividing by 10 to get $t^2 + 3t - 20 = 0$. (Review this from the Year 3 unit *Fireworks* as needed.) Students will probably solve the equation using a guess-and-check approach. To the nearest tenth of a second, the relevant solution is 3.2 seconds.

> Students may realize, especially if they consider the graph of the function $h = t^2 + 3t - 20$, that the equation $t^2 + 3t - 20 = 0$ has two solutions, $x \approx 3.2$ (which represents the solution to the pillow problem) and $x \approx -6.2$. If this second solution comes up, you can ask students to explain its significance in the situation. But if no one has any clear ideas, we suggest that you let it go for now. This will come up again in the discussion of *Homework 23: Using Your ABC's*.

"How can you solve this without using guess-and-check?"

Ask students how they could solve the equation other than using guess-and-check. If no one suggests using graphing or completing the square, we suggest that you mention both of these approaches. You might have students look at solving the equation graphically now, and tell them that they will see how to get the exact solution tomorrow by completing the square.

• Question 5

Students should see that the pillow's height t seconds after Maxine sees it can be given by the equation

$$h = 200 - 30t - 10t^2$$

Post this for comparison on Day 23 with similar equations.

One goal of this question is for students to see the role played by the initial height, the initial speed, and the gravitational acceleration in determining the height of a falling object.

To bring this out, ask the class to identify where the specific numbers come from in the expression $200 - 30t - 10t^2$. Students should be able to identify the number 200 as the pillow's initial height and the number 30 as the pillow's initial speed. It may be more difficult for them to state where the coefficient 10 comes from. Have them retrace their steps to see that it is half of 20, which is the rate of acceleration.

Save the expression $200 - 30t - 10t^2$ for comparison with other, similar expressions. On Day 24, students will develop a general formula for a falling object's height after t seconds.

- ## The problem of air resistance

  As noted previously, the pillow will meet substantial air resistance. In fact, this resistance increases as the pillow goes faster, so the rate of acceleration will not be constant. You may want to point out that the mathematical model described in the problem is actually not a good one.

- ## Another approach

  Another method for solving this problem involves visualizing the pillow as having been dropped from rest from some higher altitude. One can determine, based on the acceleration rate, how long the pillow has been falling when Maxine sees it and then figure out how high it was when it was dropped.

Homework 22: Big Push

This problem is basically a Ferris wheel version of *Look Out Below!*

Katie Davidow analyzes what effect the motion of the Ferris wheel will have on the diver's path upon release.

Big Push

Earlier in this unit, you solved the Ferris wheel problem based on the simplifying assumption that once the diver was released, he would fall as if the Ferris wheel had not been moving. But that was a simplification. In this assignment, you will examine in a specific case how the motion of the Ferris wheel would affect the diver's fall.

Here is a summary of some important facts about the Ferris wheel and the cart.

• The Ferris wheel has a radius of 50 feet.

• The center of the Ferris wheel is 65 feet above the ground.

• The Ferris wheel turns counterclockwise at a constant rate, making a complete turn every 40 seconds.

• The water level in the cart is 8 feet above the ground.

Imagine that the assistant lets go of the diver at the 9 o'clock position. Because the platform is moving downward at that moment, the diver will also be moving downward as he is released. His initial speed when released is equal to the speed with which he was moving when he was on the platform, which you already have found to be 2.5π feet per second, or approximately 7.85 feet per second.

Once released, the diver's speed will increase, just as with any object falling freely, so that his speed increases by 32 feet per second for each second that he falls. For now, assume that the tub of water is in a fixed position, directly below the diver's point of release.

Continued on next page

1. What is the diver's height t seconds after he is released?

2. How long will it take from the time the diver is released until he reaches the water?

3. How long would it take the diver to fall to the water level if the Ferris wheel was not moving and the diver was simply dropped, with no initial speed, from the 9 o'clock position?

4. Compare the answers to Questions 2 and 3, and determine how far the cart would travel during a time interval equal to the difference between those answers.

Students learn about the quadratic formula.

The General Quadratic Equation

Mathematical Topics

- Describing the position of a falling object by a quadratic expression
- Solving a quadratic equation by completing the square
- Using the quadratic formula

Outline of the Day

In Class

1. Discuss *Homework 22: Big Push*

- Bring out the similarity between this problem and *Look Out Below!*

2. Discuss quadratic equations and falling objects

- Bring out that the height function for a falling object will always involve a quadratic expression
- Have students solve the equation from *Look Out Below!* by completing the square

3. Introduce the quadratic formula

- Bring out that this formula is derived by applying the method of completing the square to the general quadratic equation

- Illustrate the use of the quadratic formula by solving the equation from *Look Out Below!*

4. *Finding with the Formula*

- Students practice the use of the quadratic formula

5. Discuss *Finding with the Formula*

- Pay particular attention to issues of sign

At Home

Homework 23: Using Your ABC's

Discuss With Your Colleagues

The Quadratic Formula

The quadratic formula is something of an icon in high school mathematics. For some people, it represents the essence of "algebra."

What do you and your colleagues think? How do students react to the presentation here of the formula without a proof? (They will develop a proof as part of the subsequent unit *Know How*.) How much does their work in the Year 3 unit *Fireworks* help in the discussion here?

1. Discussion of *Homework 22:*
Big Push

Let a volunteer present Question 1. The analysis is essentially the same as that of *Look Out Below!*, except that the initial height is 65 feet instead of 200 feet, the initial speed is 7.85 ft/sec (or 2.5π ft/sec) instead of 30 ft/sec, and the acceleration is 32 ft/sec per second instead of 20 ft/sec per second. Students should find that the height after t seconds is given by the equation

$$h = 65 - 7.85t - 16t^2$$

You should save this equation for comparison tomorrow with the formulas from *Look Out Below!* and from Question 2c of tonight's homework.

For Question 2, students need to take into account that the water level is 8 feet off the ground. They should come up with an equation that is equivalent to $65 - 7.85t - 16t^2 = 8$ and should find (probably by guess-and-check) that it takes approximately 1.66 seconds for the diver to reach the water. (As with yesterday's discussion of *Look Out Below!*, you might discuss the fact that the equation has another solution, $t \approx -2.15$.)

• *Questions 3 and 4*

Students should see that if the diver were dropped from the 9 o'clock position of a stationary Ferris wheel, it would take $\sqrt{\frac{57}{16}} \approx 1.89$ seconds for him to reach the water. Therefore, the initial motion from the turning Ferris wheel shortens the time it takes him to reach the water by about 0.23 seconds.

"How far would the cart move in this amount of time?"

Question 4 asks how far the cart would move in that amount of time (if it were moving). At 15 ft/sec, it would travel about $0.23 \cdot 15 \approx 3.4$ feet, so failure to take the diver's initial speed into account might cost the diver his life!

Note: This may lead to a discussion of how big the tub of water is—that is, of how much margin of error there is in the cart's position. We don't answer that question, and you can, if you wish, let the class make a decision on this. (Students may already have discussed this issue in connection with their solution of *Moving Cart, Turning Ferris Wheel*.)

2. Quadratic Equations

Bring out that the last two problems, *Look Out Below!* and *Homework 22: Big Push,* each involved finding the solutions to a quadratic equation. Get students to state those two equations and put them on the board or a transparency for today's discussion purposes. It isn't essential that the equations be in standard form. For example, the equations might look like this.

$$10t^2 + 30t = 200$$

$$65 - 7.85t - 16t^2 = 8$$

"Why do these problems lead to quadratic equations?"

Ask why every problem of this type will produce a quadratic equation. Students will probably see that the average speed will be a linear function of *t*, so multiplying the average speed by the length of the time interval will give a quadratic expression.

• The general linear equation

Tell the class that they will be learning a general formula for solving quadratic equations based on the equation's coefficients. To understand what this means, they will look briefly at the much simpler case of linear equations.

"What are some examples of linear equations?"

Ask students for some examples of linear equations. Elicit examples as simple as $3x + 4 = 0$ and as complex as $4x - 5 = -2x + 7$.

"How can you solve this equation algebraically?"

Then, using a simple example such as $3x + 4 = 0$, ask how students might solve this equation using algebra. Although students may resist, have them go through the formal steps of subtracting 4 from both sides and then dividing both sides by 3.

Tell them that the equation $ax + b = 0$ (with $a \neq 0$) is called the **general linear equation.** Clarify that this equation is called *general* because every linear equation can be put in that form.

Have students take various linear equations and see how they fit into this form. For instance, if they start with the linear equation $3x = 7$, they can rewrite it as $3x - 7 = 0$ to see that it corresponds to the case $a = 3$ and $b = -7$. If their examples of linear equations included more complex equations, then have them find the values of *a* and *b* for one or two of those as well.

"How can you express the solution to ax + b = 0 in terms of a and b?"

Next, have groups work on finding the solution to the general linear equation. They should see that the solution can be expressed in terms of *a* and *b* as $x = \frac{-b}{a}$. (They will probably get this solution by first subtracting *b* from both sides to get $ax = -b$ and then dividing both sides by *a* to get $x = \frac{-b}{a}$.)

Bring out that what is going on here is a sequence of *equivalent equations.* That is, any value of *x* that fits any one of these equations ($ax + b = 0$, $ax = -b$, or $x = \frac{-b}{a}$) will fit the others. (We think of the last of these, $x = \frac{-b}{a}$, as "the solution" because it tells us explicitly what value of *x* we want.)

Have students apply this formula to specific examples. For instance, they should see that to solve the equation $5x - 8 = 0$, they should substitute 5 for a and -8 for b, which gives the expression $\frac{-(-8)}{5}$, which is the same as $\frac{8}{5}$. They should verify that $x = \frac{8}{5}$ gives the solution to the equation $5x - 8 = 0$.

Identify the equation $x = \frac{-b}{a}$ as giving the solution to the general linear equation.

- *Solving a specific quadratic equation by completing the square*

Tell students that a similar, but more complicated, process will work for solving the general quadratic equation. However, before they look at the general case, they should work through a specific example in detail.

"How could you solve $x^2 + 8x - 12 = 0$ using the method of completing the square?"

You might start with the equation $x^2 + 8x - 12 = 0$, asking students to solve this using the method of completing the square. As needed, review the method, which students saw in the Year 3 unit *Fireworks*.

Students might use a sequence of equivalent equations like these.

$$x^2 + 8x - 12 = 0$$
$$x^2 + 8x = 12$$
$$x^2 + 8x + 16 = 12 + 16$$
$$(x + 4)^2 = 28$$
$$x + 4 = \sqrt{28} \quad \text{or} \quad -\sqrt{28}$$
$$x = \sqrt{28} - 4 \quad \text{or} \quad -\sqrt{28} - 4$$

"How many solutions does this equation have?"

Bring out explicitly that the equation has two solutions and that this method gives the two solutions *exactly*. You might have students substitute one or both of these expressions into the expression $x^2 + 8x - 12$ to see that they give 0. (You might have students expand the general expression $(a - b)^2$ as a first step.)

You also might have students get the approximate decimal values for each of the solutions (roughly, $x = 1.3$ and $x = -9.3$). They can then substitute these values into the expression $x^2 + 8x - 12$ to see that both give approximately 0, but not exactly 0.

Introduce the symbol $\pm$ as a shorthand for "plus or minus," and show how it can be used in the context of the example just discussed. For instance, show students that the two exact solutions can be represented by the expression $\pm\sqrt{28} - 4$.

In *Fireworks,* students saw that quadratic equations often have two roots, but the symbol $\pm$ was not introduced.

3. The Quadratic Formula

Try to get students to agree that the process of completing the square could also be used on the equation $65 - 7.85t - 16t^2 = 8$ (or whatever form you used for the equation from *Homework 22: Big Push*), although you should certainly acknowledge that the algebra would be much messier. Bring out that the final solution depends on the coefficients in the equation (including the term 8 on the right side of the equation).

"What does the general quadratic equation look like?"

Have students state what the general quadratic equation might look like. If needed, review that the standard form is $ax^2 + bx + c = 0$. Tell them that it is also standard to set this equation up so that a is positive. (Point out that if $a = 0$, then we really have a linear equation, not a quadratic equation.)

Take a minute now to have students put the equation from *Homework 22: Big Push* into this standard form. They should be able to rewrite it as

$$16t^2 + 7.85t - 57 = 0$$

Tell them that the method of completing the square, used to get the *exact* solution to the equation for *Look Out Below!,* can be applied to any quadratic equation, by working with the variable coefficients a, b, and c instead of with specific numerical values. Point out that the resulting solution will, of course, be expressed in terms of these variables.

With that introduction, post the formula.

If $ax^2 + bx + c = 0$, and $a \neq 0$,

then $x = \dfrac{-b \pm \sqrt{b^2 - 4ac}}{2a}$.

"What does '±' mean here?"

For clarity, ask students what the symbol $\pm$ means in this formula. Help them to see that as in the specific example discussed previously (for the equation from *Look Out Below!*), it means that there are two solutions, one with a plus sign and one with a minus sign. That is, one solution is $\dfrac{-b + \sqrt{b^2 - 4ac}}{2a}$ and the other is $\dfrac{-b - \sqrt{b^2 - 4ac}}{2a}$.

Tell the class that this expression giving the solution in terms of the coefficients is called the **quadratic formula,** that it is derived by completing the square with the general quadratic equation, and that the derivation of this formula is part of the Year 4 unit *Know How*.

• Applying the formula

Have students apply the formula to the example worked through earlier using the process of completing the square. That is, have them identify a, b, and c for the equation $x^2 + 8x - 12 = 0$, substitute these values into the equation, and verify that it gives the same exact answers, in terms of square roots, that they found earlier.

This task involves recognizing that in the equation $x^2 + 8x - 12 = 0$, the coefficient of x^2 is 1. It also involves seeing that the expression that results from applying the quadratic formula, namely, $\frac{-8 \pm \sqrt{8^2 - 4 \cdot 1 \cdot (-12)}}{2}$, can be simplified to $\frac{-8 \pm \sqrt{112}}{2}$, and that this is equivalent to the expression found earlier. (We gave the solutions as $\pm\sqrt{28} - 4$.) In particular, this requires seeing that $112 = 2\sqrt{28}$.

4. *Finding with the Formula*

After this introduction, have groups work on *Finding with the Formula*. You may want to caution students to be careful about the use of parentheses if they enter complicated expressions in their calculators.

5. Discussion of *Finding with the Formula*

On Question 1, students should come up with both $x = -4$ and $x = 7$ as solutions. Go over the details carefully, clarifying issues about signs and how to deal with the $\pm$ symbol. For instance, bring out that the coefficient b is -3, so the term $-b$ in the numerator is "positive three." (Students often think of $-b$ as "negative b" rather than as "the opposite of b" and may think that if b is already negative, then there is no need to change the sign.)

The expression inside the square-root sign comes out to 121 in Question 1, so students can eliminate the square-root sign by recognizing that $\sqrt{121} = 11$. They still need to deal with the expression $\frac{3 \pm 11}{2}$. Help them see that they can simply write the two separate expressions, $\frac{3 + 11}{2}$ and $\frac{3 - 11}{2}$, and then simplify each of them to get 7 and -4.

Have them verify that both values actually fit the equation. (You might want to point out that this equation can be solved by factoring as well.)

Also have students graph the function given by the equation $y = x^2 - 3x - 28$ and interpret the two solutions to the equation in terms of the graph. They should see that the graph crosses the x-axis at two places, $x = -4$ and $x = 7$, which are the points on the graph where $y = 0$.

• *Question 2*

For Question 2, students will need to put the equation in standard form. They should then be able to see that they have $a = 3$, $b = 7$, and $c = -5$. According to the quadratic formula, there are two solutions, which are given by the expression $\frac{-7 \pm \sqrt{109}}{6}$. As with Question 1, go over the details. This problem involves a value of a different from 1 and a different sign for b, so it may pose issues that did not arise in Question 1.

Tell students they will learn more about quadratic equations and their solutions in the homework. Tell them also that in a subsequent Year 4 unit, *Know How,* they will examine where the quadratic formula comes from.

Note: In Question 1d of tonight's homework, students will consider a case in which the expression under the square-root sign comes out negative. The discussion of this issue tomorrow is quite brief, but the topic of complex numbers is considered further as part of *Know How*.

• *Optional: Higher degree polynomial equations*

Point out that the general quadratic equation is a particular example of a more general category called a *polynomial equation*. (The term *polynomial*, and related terms like *degree* and *cubic*, were introduced in the Year 3 unit *Fireworks,* but you may need to review them here.)

Review that the general linear equation looks like $ax + b = 0$ (with $a \neq 0$). Have students write the general cubic equation to illustrate the way the idea extends to higher degrees.

Tell students that mathematicians knew for many years how to solve linear and quadratic equations and devoted great energy and effort looking for formulas for polynomial equations of higher degree than 2.

You might mention that general formulas for solving cubic (degree 3) equations and quartic (degree 4) equations were first developed in the sixteenth century and that in the nineteenth century, it was proved that there is no similar general formula for equations of degree 5 or higher.

You might mention as well that the proof of this remarkable result uses the theory of permutations, which is related to the counting problems that students studied in the Year 3 unit *Pennant Fever.* (The proof is usually done as part of the study of Galois theory, which is an area of mathematics usually not taught until graduate school.)

Homework 23: *Using Your ABC's*

The purpose of this assignment is to get students somewhat more comfortable with the quadratic formula.

Question 1d introduces them to the issue of the nonexistence of real solutions.

Finding with the Formula

Sometimes, when repeating the same process over and over, it becomes easier simply to develop a formula that gives you the same result.

Solving quadratic equations is one of those situations. You can use the method of completing the square for individual examples, but there is a general formula that saves the trouble of repeating the steps each time.

The **general quadratic equation** is usually written in the form $ax^2 + bx + c = 0$, where the coefficients a, b, and c can be any numbers except that a cannot be 0. If you apply the method of completing the square to the general quadratic equation, you get an expression, called the **quadratic formula,** that gives the solutions in terms of the coefficients a, b, and c. The general result says

If $ax^2 + bx + c = 0$, and $a \neq 0$,

then $x = \dfrac{-b \pm \sqrt{b^2 - 4ac}}{2a}$

That is, if x is a solution to the equation $ax^2 + bx + c = 0$, then x must be equal to either $\dfrac{-b + \sqrt{b^2 - 4ac}}{2a}$ or $\dfrac{-b - \sqrt{b^2 - 4ac}}{2a}$. (You will develop a proof for this formula as part of the unit *Know How* later in Year 4.)

1. Use the quadratic formula to solve the equation $x^2 - 3x - 28 = 0$, and check your answers.

2. Use the quadratic formula to solve the equation $3x^2 + 7x = 5$, and check your answers.

Using Your ABC's

1. Find all the solutions to each of these equations using the quadratic formula. Give exact solutions, using square roots if necessary. Also approximate the solutions to the nearest tenth, and use the equations to confirm that these solutions seem correct.

   a. $x^2 + 7x + 12 = 0$

   b. $x^2 - 3x - 8 = 0$

   c. $2x^2 + 5x - 1 = 0$

   d. $2x^2 - 3x + 4 = 0$

2. Set up and solve quadratic equations to answer each of these questions.

   a. A rectangle with one side 5 feet longer than the other has an area of 126 square feet. What are the dimensions of the rectangle?

   b. A right triangle has one leg 6 inches shorter than the other, and its hypotenuse is 13 inches long. How long are the legs of the right triangle?

   c. An object is thrown up into the air off the roof of a building so that its height h in feet after t seconds is given by the equation $h = 90 + 50t - 16t^2$. When will the object be 120 feet high?

DAY 24

Three O'Clock Drop

Mathematical Topics

- Using the quadratic formula to solve problems
- Examining the significance of having two solutions to quadratic questions
- Considering the sign of velocity and its relationship to acceleration
- Finding an expression for the position of a falling object with initial upward velocity as a function of time

Outline of the Day

In Class

1. Discuss *Homework 23: Using Your ABC's*

- Bring out that the quadratic formula shows that some quadratic equations have no solution
- Consider possible interpretations of the two solutions to quadratic equations arising from problem situations

2. Introduce the distinction between velocity and speed

- Establish the convention that positive velocity in the vertical direction means upward motion
- Clarify how gravitational acceleration affects velocity in the context of changing signs

3. *Three O'Clock Drop*

- Students examine a Ferris wheel situation in which the diver has initial upward motion
- The activity will be discussed on Day 25

At Home

Homework 24: Up, Down, Splat!

---

1. Discussion of *Homework 23:*
Using Your ABC's

You might have each group prepare a transparency for a different part of Question 1 or Question 2, and then have heart card students make presentations.

- ## Questions 1a through 1c

    Question 1a has integer solutions ($x = -3$ and $x = -4$), and the presenter should be able to describe how he or she used the quadratic formula to get them both.

    Questions 1b and 1c are a bit more difficult, because the exact solutions require square roots. You may need to caution students to pay particular attention to signs. Here are the solutions.

    - For Question 1b: $\frac{3 + \sqrt{41}}{2}$ and $\frac{3 - \sqrt{41}}{2}$, which are approximately 4.7 and -1.7

    - For Question 1c: $\frac{-5 + \sqrt{33}}{4}$ and $\frac{-5 - \sqrt{33}}{4}$, which are approximately 0.19 and -2.7

- ## Question 1d

    Question 1d presents a new challenge, because the equation has no real roots. Students should have found that the quadratic formula yields the expression $\frac{3 \pm \sqrt{-23}}{4}$.

    "What does this expression mean?"
    Ask students what this expression means. Be sure they articulate that there is no number (that they know of) whose square is -23 and that this means the equation has no solutions (within the number system as they probably know it so far).

    "What does this mean in terms of the graph of $y = 2x^2 - 3x + 4$?"
    Also ask what this lack of a solution means in terms of the graph of the equation $y = 2x^2 - 3x + 4$. Students should be able to articulate that the graph has no x-intercepts. Have them give a rough sketch of what such a graph might look like. (You need not get details here—merely an example of a parabola opening upward that is entirely above the x-axis.) They can verify this graph using their graphing calculators.

    Tell students that in the Year 4 unit *Know How,* they will learn about a system of numbers called the **complex numbers**—a system that includes more than the numbers represented on the number line—in which negative numbers have square roots.

- ## Question 2

    For the different parts of Question 2, keep in mind that students may make different choices about what their variables represent, and this may lead to different equations. Have presenters show how to use the quadratic formula in each case to get exact solutions. (You need not belabor the details, but if there seem to be common errors, take time to discuss them.) Be sure in each case to talk about what the solution to the equation means in terms of the problem.

On Question 2a, if students use x to represent the length of the shorter side, this leads to the equation $x(x + 5) = 126$. This is equivalent to $x^2 + 5x - 126 = 0$, which has $x = 9$ and $x = -14$ as its solutions. If $x = 9$, then the longer side of the rectangle is 14, and the rectangle is 9 feet by 14 feet.

Students should recognize that $x = -14$ does not make literal sense as the length of the side of a rectangle, but they may notice that this solution has the same absolute value as the longer side of the rectangle. Students might interpret this solution as representing a rectangle with sides whose "lengths" are -14 feet and -9 feet. (Such a rectangle could be interpreted as lying in the third quadrant of a coordinate system, in which both x and y are negative.)

Comment: Students might set up their equations using the variable to represent the longer side of the rectangle. If so, then the equation and its solutions will be different, but the dimensions of the rectangle will be the same.

On Question 2b, if x represents the longer leg, then the equation becomes $x^2 + (x - 6)^2 = 13^2$. In standard form, this becomes $2x^2 - 12x - 133 = 0$, and students can use the quadratic formula to get the two solutions $\frac{12 \pm \sqrt{1208}}{4}$, which yields values of approximately $x = -5.7$ and $x = 11.7$. Again, only one solution makes literal sense, and the triangle has legs of length 11.7 and 5.7. (As with Question 2a, students might give an interpretation to the negative solution and might set up the equation differently.)

On Question 2c, students should get the equation simply by setting the expression $90 + 50t - 16t^2$ equal to 120. In standard form, the equation is $16t^2 - 50t + 30 = 0$, which has two solutions, $\frac{50 \pm \sqrt{580}}{32}$. This yields values of approximately $t = 0.81$ and $t = 2.3$.

"What does the equation $h = 90 + 50t - 16t^2$ mean in terms of the situation?"

Before discussing whether both solutions make sense, you should ask students what the equation $h = 90 + 50t - 16t^2$ means in terms of the situation. Students will probably be able to interpret this as meaning that the object started with a height of 90 feet and an upward speed of 50 feet per second.

"Which of the solutions to the equation make sense in the problem?"

With this in mind, ask which of the solutions to the equation make sense. One solution ($t = 0.81$) indicates the time when the object reaches 120 feet on the way up, and the other solution ($t = 2.3$) indicates the time when the object reaches 120 feet on the way down.

- *Interpreting the negative solution to "Look Out Below!"*

   After students have interpreted the two solutions to Question 2c in terms of the object going up and then down, you may want to go back to *Look Out Below!*, in which students saw that the equation they used had two solutions, $t = 3.2$ and $t = -6.2$. If they did not come up with an interpretation then for the negative solution, you might see if it makes more sense to them now.

   They might imagine that someone threw the pillow upward with just the right force so that it would pass Maxine on the way down at a speed of 30 feet per second. The solution $t = -6.2$ means that the pillow would have to have been thrown upward 6.2 seconds before Maxine saw it. (As noted earlier, the issue of air resistance makes the mathematical model for this problem a poor one. That issue may also make this interpretation seem somewhat unrealistic.)

- *The general height formula*

   Ask students to develop a general formula for the height of a freely falling or rising object in terms of its initial height and initial velocity. They should come up with a principle like the one stated here, which generalizes the similar statement from Day 10, and you should post this new principle.

   Suppose a freely falling or rising object has an initial height of h feet and an initial velocity of v feet per second (where a positive value for v means upward motion). Then its height after t seconds is $h + vt - 16t^2$ feet.

- *The case $v = 0$*

   Ask students to compare the statement just developed with the principle posted on Day 10 giving the expression $h - 16t^2$ for the height of an object falling *from rest* from a height of h feet. They should see that the "from-rest" formula, for the case $v = 0$, is a special case of the general formula. (Remind them, if needed, that the initial velocity of the object in *Free Fall* was zero.)

2. Velocity Is Speed with Direction

Question 2c of the homework provides a good opportunity to talk about the difference between speed and velocity.

Review (as discussed earlier) that in Question 2c, the object's initial height was 90 feet and that it had an initial upward speed of 50 feet per second. Then have the class compare the equations for height for the last several problems.

- $h = 90 + 50t - 16t^2$ for Question 2c of last night's assignment

- $h = 200 - 30t - 10t^2$ for *Look Out Below!*

- $h = 65 - 7.85t - 16t^2$ for *Homework 22: Big Push*

Bring out that in Question 2c of last night's homework, the coefficient of t is positive, while in the other problems, the coefficient of t is negative. Ask what the significance is of the sign of the coefficient of t. Students will probably be able to explain that in last night's problem, the object was going up to begin with, and in the other problems, the object was going down initially.

Point out that in the Ferris wheel situation, there will be some release points that yield an initial upward motion and others that yield an initial downward motion. Therefore, it is important to develop a convention for distinguishing between the two.

• Velocity has sign

Tell students that **velocity** is the term used in physics to indicate the combination of speed and direction. That is, the velocity of an object tells both the speed at which the object is traveling and the direction of the motion. Velocity may be positive or negative, but speed is never negative. Speed, in fact, is the absolute value of velocity.

Bring out that until last night's homework, there was no need to worry about this, because the situations students were considering all involved objects that were moving downward. But if students are going to consider objects that go up and then down, they need to distinguish between, say, an upward speed of 50 ft/sec and a downward speed of 50 ft/sec.

In the context of vertical motion, it usually makes sense to think of "up" as the positive direction. Tell students that they should use this convention for the remainder of the unit. (You can acknowledge that this is a somewhat arbitrary choice.)

With this convention in mind, go back to the three problems to see why the coefficient of the linear term in the height function has the sign it does in each case. Help students reach these conclusions.

- In *Look Out Below!*, the pillow had a velocity of -30 feet per second when Maxine first saw it.

- In *Homework 22: Big Push,* the diver had a velocity of -7.85 feet per second when he was first released from the Ferris wheel.

- In Question 2c of *Homework 23: Using Your ABC's,* the object had an initial velocity of 50 feet per second.

• Acceleration and the sign of velocity

There is potential for confusion about the idea of constant acceleration in a context in which the sign of the velocity is changing. We have established the convention that a positive velocity represents upward motion. While an object is moving upward, the effect of gravity is to slow it down, which means to decrease both its speed and its velocity. On the other hand, when an object is moving downward, the effect of gravity is to increase its speed.

You may want to clarify that in both cases, the *velocity* is decreasing, in the first case going from positive to "less positive" and in the second case going from negative to "more negative." Tell students that the rate at which the speed of a rising object decreases is the same as the rate at which the speed of a falling object increases. Bring out that based on the convention for the sign of velocity, the acceleration due to gravity has a negative sign, which is reflected in the sign of the term $-16t^2$ in expressions for the height of falling objects.

You might illustrate the effect of gravitational acceleration with an example involving a change of sign of velocity. For instance, have students imagine an object with an initial velocity of 50 ft/sec (which represents upward motion), and ask what the object's speed will be after 1 second and after 2 seconds. They should see during the first second, the velocity goes from 50 ft/sec to 18 ft/sec, and during the next second, it goes from 18 ft/sec to -14 ft/sec. Bring out that at each stage, the velocity decreases by 32 ft/sec, whether it was initially positive or negative.

3. Three O'Clock Drop

This activity is almost identical to *Homework 22: Big Push*. The only difference is that in this activity, the initial velocity is upward instead of downward, and no special introduction is needed. This activity is scheduled to be discussed tomorrow.

Homework 24: Up, Down, Splat!

This assignment puts initial upward motion in a different context.

Three O'Clock Drop

In *Homework 22: Big Push,* you found the falling time for the diver if he was released from the 9 o'clock position. In that problem, the motion of the Ferris wheel gave him an initial downward velocity as he was released, and you saw that it took him less time to fall than if he had been dropped from a stationary Ferris wheel.

Now consider what happens if he is released from the 3 o'clock position. Because the platform is moving upward at that moment, the diver will start off with an initial upward motion. As in *Homework 22: Big Push,* his initial speed is equal to the speed with which he was moving when he was on the platform.

Assume again that the cart is in a fixed position, directly below the diver's point of release.

1. How long will it take from the time the diver is released until he reaches the water?

2. How long would it have taken him to reach the water if he had been released from a motionless Ferris wheel (at the 3 o'clock position)?

Up, Down, Splat!

Melissa's science class is having a contest. The contest is to see who can build a container that will keep an egg from breaking when dropped from the school window.

Melissa is quite confident of her contraption. She leans out the window, which is 25 feet off the ground, and hurls her egg container straight up in the air with an initial velocity of 35 feet per second. (Consider velocity upward to be positive.)

Assume that the egg container's velocity is affected by gravity in the usual way. That is, the velocity decreases by 32 feet per second for each second the egg container travels.

1. How long does it take for the egg container to hit the ground?

2. At what speed does the egg container hit the ground?

Objects Going Up and Down

Mathematical Topics

- Consolidating ideas about falling objects with nonzero initial velocity

Outline of the Day

In Class

1. Discuss *Three O'Clock Drop* (from Day 24)

- Compare the analysis to that of *Homework 22: Big Push*

2. Discuss *Homework 24: Up, Down, Splat!*

- Find the egg container's speed on impact

At Home

Homework 25: Falling Time for Vertical Motion

Note: We suggest that you discuss yesterday's activity, *Three O'Clock Drop,* before discussing *Homework 24: Up, Down, Splat!*

1. Discussion of *Three O'Clock Drop*

For Question 1, students will probably develop an equation for the time the diver is falling from a formula for the object's height in terms of time. They might simply write down this formula, based on earlier examples, without going through the analysis of finding the average velocity. (You may decide to review the "averaging the endpoints" method, as it was used in *Look Out Below!,* for example, but that will depend on your class.)

*"Why is this
expression for
the diver's height
almost exactly like
that in 'Homework 22:
Big Push'?"*

The diver's height t seconds after release is given by the expression $65 + 7.85t - 16t^2$. As a follow-up to yesterday's discussion of the sign convention for velocity, ask why this expression is identical to that for *Homework 22: Big Push* except for the sign of the coefficient of t. Students should see that the diver is being released from the Ferris wheel at the same height and with the same speed in both cases. The distinction is that in *Three O'Clock Drop*, his initial motion is upward, while in *Homework 22: Big Push*, his initial motion was downward.

To answer Question 1, students need to determine when the diver is 8 feet off the ground, so they need to solve the equation $65 + 7.85t - 16t^2 = 8$. The solutions to the equation are $t \approx 2.15$ seconds and $t \approx -1.66$ seconds. (Encourage students to use the quadratic formula to get the exact values in terms of square roots. In tonight's *Homework 25: Falling Time for Vertical Motion*, they will need to use the formula to get a general expression for the falling time for a free-falling object.)

You might discuss once again that the equation only imperfectly represents the problem, because only one of its solutions makes sense in this context.

Next, let a volunteer present Question 2. The presenter might simply substitute 57 for h in the expression $\sqrt{\frac{h}{16}}$ (using the formula developed on Day 10 and stated in *Homework 10: A Practice Jump*). Or he or she might point out that the time is the same as that for Question 3 of *Homework 22: Big Push*, namely, $t \approx 1.89$ seconds. You can bring out that students can also find this value by solving the equation obtained by eliminating the term $7.85t$ from the equation used in Question 1, that is, by solving $65 - 16t^2 = 8$.

*"How much of a
difference does the
diver's upward motion
from the movement of
the Ferris wheel make
in his falling time?"*

Next, ask how much of a difference the diver's initial upward motion from the movement of the Ferris wheel makes in the time needed for the diver to reach the water level. Students can calculate that it takes approximately an additional 0.26 seconds. (You might compare this to the decrease in falling time resulting from the effect of the Ferris wheel's motion in *Homework 22: Big Push*, which was approximately 0.23 seconds. Although the amounts are roughly the same, they are not identical.)

2. Discussion of *Homework 24: Up, Down, Splat!*

Question 1 involves essentially the same process as in the discussion of *Three O'Clock Drop*, but with different numbers. Here, the height is given by the expression $25 + 35t - 16t^2$, and the solutions to the equation $25 + 35t - 16t^2 = 0$ are $t \approx 2.75$ and $t \approx -0.57$. Students presumably will see that the solution $t \approx 2.75$ is the one they want in this problem.

You might take this opportunity to make sure students are connecting their algebra to the situation by asking what the number 2.75 represents. They should see that it gives the number of seconds from when Melissa throws the egg container upward until it hits the ground.

• Question 2: The speed at impact

Question 2 represents an issue that has not yet been discussed in this sequence of problems. The simplest approach is to use the fact that the velocity is decreasing by 32 feet per second for each second, so the velocity after t seconds is given by the expression $35 - 32t$. Students can then substitute $t \approx 2.75$ seconds into this expression to get a value of about -53 feet per second for the velocity when the egg container hits the ground. Be sure to discuss the significance of the negative sign.

Homework 25: Falling Time for Vertical Motion

This assignment represents a culmination of the last several activities. It asks students to develop a general expression for the falling time of an object in terms of its initial velocity and the distance it needs to fall. (We have suggested that students rewrite the equation in Question 1 with a positive x^2-coefficient so that the final expression is easier to work with.)

Falling Time for Vertical Motion

In each of several recent problems, you figured out how long it took for an object to fall a certain distance if it started with a certain initial velocity.

When an object is falling freely, its height after t seconds is given by the expression $h + vt - 16t^2$, where h is the object's initial height and v is the object's initial velocity (where upward motion is considered positive). Finding out when the object hits the ground is equivalent to solving the equation $h + vt - 16t^2 = 0$.

In specific cases, you might be able to solve the equation (or get a good estimate) by guess-and-check or with a graph. But when you solve the main unit problem, you will not have numerical values for h or v, because those coefficients will be expressed in terms of the variable W. In preparation for dealing with that complication, your task here is to solve the equation $h + vt - 16t^2 = 0$ *in terms of h and v.*

1. Rewrite the equation $h + vt - 16t^2 = 0$ so that it is in the form $ax^2 + bx + c = 0$, with a positive value for a.

2. Use the quadratic formula to solve the equation from Question 1. Your answer should give t in terms of h and v.

3. Which of the two solutions you found in Question 2 will give you a positive value for t? (*Hint:* Assume that h is a positive number.)

Components of Velocity

This page in the student book introduces Days 26 through 30.

Stephanie Skangos and DeAnna DelCarlo consider what other directions are possible for the diver upon release from a moving Ferris wheel.

In several recent problems, you've considered objects with either positive or negative initial *vertical* velocity. But when the Ferris wheel diver is released by his assistant, he might go sideways as well as up or down.

How do you take this into consideration? How do the vertical and horizontal parts of his motion work together? How does gravity fit into the picture? These are the sorts of questions you need to answer next.

DAY 26

High Noon

Mathematical Topics

Students consider a case in which the diver has an initial horizontal motion.

- Using the quadratic formula to get a general expression for the falling time of an object in terms of initial height and velocity
- Relating vertical and horizontal motion
- Studying the motion of falling objects when the horizontal component of initial velocity is nonzero

Outline of the Day

In Class

1. Discuss *Homework 25: Falling Time for Vertical Motion*

- Post the formula for the falling time in terms of initial height and velocity
- Discuss what's next in the main unit problem

2. Discuss that horizontal motion and gravitational fall work independently of each other

3. *High Noon*

- Students consider a case in which the Ferris wheel diver has an initial horizontal motion
- The activity will be discussed on Day 27

At Home

Homework 26: Leap of Faith

Special Materials Needed

- (Optional) Materials to demonstrate the independence of horizontal motion and the effect of gravity

1. Discussion of *Homework 25: Falling Time for Vertical Motion*

You can let students compare work in their groups and then have volunteers make presentations.

For Questions 1 and 2, students should rewrite the equation, probably as $16t^2 - vt - h = 0$, and then simply apply the quadratic formula, which gives

$$t = \frac{v \pm \sqrt{(-v)^2 - 4 \cdot 16 \cdot (-h)}}{32}$$

which simplifies to

$$t = \frac{v \pm \sqrt{v^2 + 64h}}{32}$$

The task in Question 3 is to decide whether to use the $+$ sign or the $-$ sign. Students may tap their experience with previous problems, or they may be able to analyze the situation. For example, because h is assumed to be positive, the expression inside the square-root sign is bigger than v^2, so using the $-$ sign would result in a negative value for t.

Post this solution, with a description of what it represents. The posted statement should say something like this:

Suppose an object is put into the air with an initial vertical velocity of v feet per second and falls a net distance of h feet. Then the time for this fall is given by the expression

$$\frac{v + \sqrt{v^2 + 64h}}{32}$$

You might bring out that this expression is smaller if v is negative than if v is positive with the same absolute value. That is, the object will reach the ground faster if it is initially moving downward.

• *What if v = 0?*

"What does this formula say about the case v = 0?"

This complex formula may intimidate students. One way to make it less mysterious, as well as to review ideas discussed previously, is to ask what the formula says about the case $v = 0$. That is, does this formula agree with what students learned earlier about objects falling from rest?

Let students investigate this for a few minutes. Their first step should be to replace v with 0, which simplifies the expression considerably. But they may not immediately connect the resulting expression, $\frac{\sqrt{64h}}{32}$, with their earlier result, $\sqrt{\frac{h}{16}}$.

"Are the expressions $\frac{\sqrt{64h}}{32}$ and $\sqrt{\frac{h}{16}}$ equivalent?"

Have groups examine whether these two expressions are equivalent. If they get stuck, you might ask whether they can write $\frac{\sqrt{64h}}{32}$ as simply a radical or whether they can write $\sqrt{\frac{h}{16}}$ with a radical in the numerator only. Another option is for them to square and then simplify both expressions, and see that the results are the same.

Note: The topic of simplifying radical expressions was discussed briefly in the Year 2 unit *Do Bees Build It Best?* (see *Homework 18: Simply Square Roots*).

• *What's next?*

"What do you think comes next in the solution of the unit problem?"

You can ask students what they think comes next in the solution of the unit problem. Point out that in the 3 o'clock and 9 o'clock positions, the diver's initial motion either shortened or lengthened the time it took for him to fall to the water level (as compared to the time it would have taken had he fallen from rest).

"What other effect might the motion of the Ferris wheel have on the diver besides shortening or lengthening the time of his fall?"

Tell students to suppose now that the diver is released at some position other than 3 o'clock or 9 o'clock. Ask what effect the turning of the Ferris wheel would have on his motion other than shortening or lengthening the time of his fall. Bring out that he would also be moving to one side or the other and that this would cause him to land in a different place than if he were dropped from rest. Tell students that the next aspect of the unit is to learn how to combine this sideways motion with vertical motion.

2. Horizontal Motion and Gravitational Fall Are Independent

Before dealing with these other cases of the unit problem, students need to learn one more principle about the physics of motion. We recommend that you consult your physics department for ideas about experiments or videos that might be helpful here.

Tell students that the key idea for combining horizontal and vertical motion is to treat them completely separately. Acknowledge that this principle may not fit their intuition but that it can be verified experimentally. You can summarize the way the principle works as follows.

If an object is moving sideways as it falls, then

- **the rate at which its height changes is the same as if it were falling straight down**

- **the rate at which it moves sideways is the same as if there were no gravity**

Bring out that the object's overall speed is a blend of the vertical and horizontal speeds and is greater than either of these separate speeds. (The term used in physics for this combined motion is *resultant*.)

3. *High Noon*

This activity examines the case of the diver being released from the 12 o'clock position and applies the ideas of the discussion just completed. No further introduction is needed.

Homework 26: Leap of Faith

This assignment applies the same ideas as needed in *High Noon*, but in a context different from the Ferris wheel.

Yaravi Anaya, Audrey Rae, and Shaleen Nand consider what other directions are possible for the diver upon release from a moving Ferris wheel.

High Noon

At the moment when the diver reaches the 12 o'clock position, his motion on the Ferris wheel is purely horizontal. So if he were released at that moment, he would not be moving up or down at all, but only to the side.

Because of this, the diver's falling time is the same as if he had fallen from a motionless Ferris wheel. But the diver will continue to move sideways throughout his fall. His sideways motion will be at the same rate as the platform was moving.

Use these facts, together with everything else you know about the Ferris wheel and the motion of falling objects, to answer these questions.

1. How long will it take for the diver to reach the water level?

2. How far to the left of center will the diver be when he is 8 feet off the ground? In other words, what is his x-coordinate when he reaches the water level?

Leap of Faith

"Fire! Fire!" someone yells from down the hall. You reach for the doorknob to look out the door, but remember to feel the door first. It's a good thing you did, because the door is extremely hot, and the fire is working its way through the door.

You are able to go out the window and make your way up to the roof, which is flat. As you contemplate your situation, the firefighters arrive.

"Jump! Jump!" they shout, and they hold out a rescue net. The net is 30 feet below you, and its center is 15 feet out from the edge of the roof.

You decide to run straight off the edge of the roof, hoping to go just far enough out so as to land in the middle of the net.

1. At what speed should you be going as you leave the roof in order to hit the center of the net? (*Hint:* First figure out how long it will take you to fall the 30 feet.)

2. If the net is 10 feet across, then what range of speeds will allow you to hit the net?

DAY 27

The Ideal Skateboard

Mathematical Topics

- Analyzing velocity in terms of vertical and horizontal components

Students learn about "components of velocity."

Outline of the Day

In Class

1. Discuss *Homework 26: Leap of Faith*
   - Bring out that the falling time can be found independently of the horizontal motion
2. Discuss *High Noon* (from Day 26)
3. Introduce the concept of components of velocity

4. *The Ideal Skateboard*
   - Students examine a situation involving both vertical and horizontal components of velocity
   - The activity will be discussed on Day 28

At Home

Homework 27: Racing the River

---

1. Discussion of *Homework 26: Leap of Faith*

Ask for a volunteer to present Question 1. The presenter should recognize, based on what the class learned yesterday, that the falling time is the same as that for an object falling 30 feet from rest, so this time is $\sqrt{\frac{30}{16}}$ seconds, or about 1.37 seconds.

Students should then apply yesterday's second principle, to see that the jumper should cover the same distance horizontally as if he or she had continued moving horizontally at a constant rate. The task is to find that rate. Because the jumper wants to go 15 feet in 1.37 seconds, the appropriate rate is $\frac{15}{1.37}$ ft/sec, or about 11 ft/sec.

Question 2 adds an interesting side issue to the problem. If the net is 10 feet across, then the jumper can go anywhere from 10 feet to 20 feet and still land in the net (although it's probably important to land somewhere near the

---

middle). Therefore, the jumper's initial speed coming off the roof can be as little as $\frac{10}{1.37}$ ft/sec (about 7.3 ft/sec) and as much as $\frac{20}{1.37}$ ft/sec (about 14.6 ft/sec).

2. Discussion of *High Noon*

You may wish to give groups more time to work on this problem in light of the homework discussion. This will give you an opportunity to determine if the presentation of this assignment will be fruitful for the whole class or will simply be a repetition of the ideas in the homework discussion.

> At the 12 o'clock position, the diver is 107 feet above the water level, so his falling time (down to the water level) is $\sqrt{\frac{107}{16}} \approx 2.59$ seconds. His horizontal speed at the moment of release is 7.85 ft/sec, so he will travel $7.85 \cdot 2.59 \approx 20.3$ feet to the left as he falls.

3. Vertical and Horizontal Components

Ask students to list the specific cases of the diver's release that they have looked at.

- The 9 o'clock position, in *Homework 22: Big Push*

- The 3 o'clock position, in *Three O'Clock Drop*

- The 12 o'clock position, in *High Noon*

"What is special about these cases?"

Ask them what they think is special about these cases. Someone will probably recognize that they are all situations in which the diver's initial velocity is either purely vertical or purely horizontal.

Bring out that in *High Noon*, although the diver's initial velocity was horizontal, his movement overall was both vertical and horizontal. Ask students to summarize how they dealt with this combination of horizontal motion with the downward force of gravity. They should be able to articulate that they treated the two parts of the diver's motion separately, using two principles.

- The diver moves sideways as if there were no gravity.

- The diver moves down as if there were no sideways motion.

Review that according to both theoretical and experimental physics, we can always treat these two parts of motion as if they were completely separate. Introduce the phrases **vertical component of velocity** and **horizontal component of velocity** for these two parts of the diver's (or any object's) motion.

"What was the vertical component of the diver's velocity 1 second after he was released? What was the horizontal component?"

Go over how this idea works in the context of *High Noon*. Ask students about the components of the diver's velocity 1 second after release. They should see that the vertical component would be the same as if he had fallen from rest, namely, −32 ft/sec. (This velocity is negative because he is moving downward.) The horizontal component doesn't change during the diver's fall, so it is the same as when he started, namely, −7.85 ft/sec. (This velocity is negative because he is moving to the left.)

Make sure students realize that for the vertical component, we need to be thinking about instantaneous velocity, because this component is changing at each second. You might point out that this change in the vertical component explains why the diver's path gets steeper as he falls.

4. *The Ideal Skateboard*

This activity provides a context for examining components of velocity without the complication of the issue of gravity.

As a hint in Question 3, you can remind students that the skateboarder's path is tangent to the circle, so it is perpendicular to the "2 o'clock" radius of the circle.

Homework 27: Racing the River

This assignment, which uses two swimmers to help identify the directional components of speed, is quite similar to *The Ideal Skateboard*.

You may want to clarify that students are to treat this situation as if there is no river current.

"Teaching High Dive *really helped me understand the components of velocity in a way that made a lot more sense than in my college physics classes."*

IMP teacher Betsy Adams

The Ideal Skateboard

Let's consider a skateboard situation like the one you looked at in *Homework 20: Initial Motion from the Ferris Wheel*. Imagine a skateboarder holding onto a spinning platform in the middle of a skateboard park.

Here are some more details.

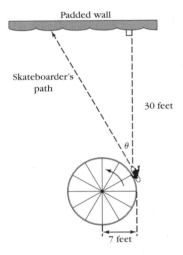

Padded wall

Skateboarder's path

30 feet

θ

7 feet

• The platform has a 7-foot radius and makes a complete turn every 6 seconds.

• The skateboarder lets go from the 2 o'clock position, as shown in the diagram.

The skateboarder will eventually crash into the padded wall. At the moment of release from the platform, the skateboarder is 30 feet from this wall.

1. How fast will the skateboarder travel? (Assume as before that there is no friction.)

2. What is the angle shown in the diagram as θ?

3. How much closer will the skateboarder be to the wall after each second? In other words, what is the "toward the wall" component of the skateboarder's velocity?

4. Use your answer to Question 3 to find out how long it takes for the skateboarder to reach the wall.

5. Find the actual distance the skateboarder travels, and use that information (and the answer to Question 1) as an alternate way to find out how long it takes for the skateboarder to reach the wall.

Racing the River

One year, as part of their River Days Festival, the citizens of River City were looking for a new event to raise money for charity.

Someone had the idea of having groups of swimmers from the two local high schools, River High and New High, compete in a race across the river. The town would raise money by charging admission to watch the race.

The river was generally quite calm during the time when the festival was held, so the planners could safely assume that the current would not affect the swimmers. The race was set to take place along a straight stretch where the river was 200 meters across.

Representatives of New High pointed out that the River City swim team was based at River High, so most of the best swimmers attended that school. To even things out, the planners decided to make the River High swimmers swim farther. The arrangement they agreed on was that the New High swimmers would swim directly across the river while the River High swimmers would swim at an angle of 45° off from the direct route across the river, as shown here.

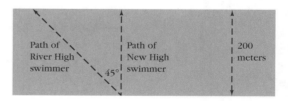

Continued on next page

1. If a River High swimmer swims at a rate of 1.5 meters per second, how long will it take the swimmer to get to the other side? (*Hint:* How far does a swimmer from River High need to swim to get across?)

2. How fast must a New High swimmer swim to get to the other side at the same time as the River High swimmer?

The next year, the planners decided that the New High swimmers did not need such a big advantage. They changed the path for the River High swimmers so it would make only a 30° angle with the direct route, as shown here.

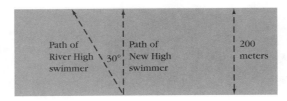

3. Assume that a River High swimmer still swims at a rate of 1.5 meters per second. Based on this new rate and new angle, answer these questions.

    a. How far does a swimmer from River High need to swim to get across?

    b. How long will it take a River High swimmer to get to the other side?

    c. How fast must a New High swimmer swim to get to the other side at the same time as the River High swimmer?

DAY
28

Students consider a Ferris wheel problem involving both vertical and horizontal initial motion.

One O'Clock Without Gravity

Mathematical Topics

- Expressing velocity in terms of vertical and horizontal components
- Studying the motion of falling objects when both the vertical and horizontal components of initial velocity are nonzero

Special Materials Needed

- A diagram for use in discussing *Homework 27: Racing the River* (see Appendix B)

Outline of the Day

In Class

1. Discuss *Homework 27: Racing the River*
- Identify the "toward-shore" component of the swimmer's motion

2. Discuss *The Ideal Skateboard* (from Day 27)
- Clarify how to determine the angle of the skateboarder's motion
- Bring out that the vertical and horizontal components of velocity work independently

3. *One O'Clock Without Gravity*
- Students consider a Ferris wheel problem involving both vertical and horizontal initial motion
- The activity will be discussed on Day 29

At Home

Homework 28: General Velocities

1. Discussion of *Homework 27: Racing the River*

The key idea is for students to recognize that the River High swimmer's speed has a "toward-shore" component that depends on the angle at which the swimmer is moving.

Questions 1 and 2 should be fairly straightforward. Students might find the distance that the River High swimmer must swim using either the Pythagorean theorem or trigonometry. This distance (approximately 283 meters) can then be divided by the swimmer's speed (1.5 m/sec) to get the time it will take for the swimmer to finish the race (approximately 189 seconds). On Question 2, students will probably divide the distance by the time to get the rate (200 m/189 sec ≈ 1.06 m/sec).

Once the rate for the New High swimmer has been determined, you can get at the idea of the River High swimmer's toward-shore component with a pair of questions like this.

- How far from the starting shore is the New High swimmer after 50 seconds?

- How far from the starting shore is the River High swimmer after 50 seconds?

The first question can be answered simply by multiplying the New High swimmer's rate (1.06 m/sec) by 50 seconds. Bring out that the second question must have the same answer as the first. Use that observation to establish that the River High swimmer is swimming toward the shore at a speed of 1.06 m/sec, even though he is actually swimming at 1.5 m/sec.

Identify this value, 1.06 m/sec, as the "toward-shore" component of the River High swimmer's speed.

- *Question 3*

    The three parts of Question 3 are essentially a repeat of Questions 1 and 2, but with a different angle. The goal here is to bring out that for a given speed, the toward-shore component of the River High swimmer's rate depends on the angle at which the swimmer is moving.

    You can use a general diagram like the one shown here to see that the toward-shore component can be found by multiplying the overall speed by cos θ, where θ is the angle between the actual path and the direct route.

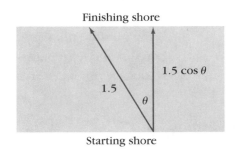

That is, for each 1.5 meters the swimmer moves at an angle θ from the direct route, he or she gets closer to shore by $1.5 \cos \theta$ meters. (You might have students consider the special cases $\theta = 0°$ and $\theta = 90°$ as confirmation of this result.)

You can illustrate this idea physically using a transparency of the diagram (see Appendix B) by placing a ruler along the starting shore and moving it gradually toward the finishing shore.

2. Discussion of *The Ideal Skateboard*

Although the principles used in *The Ideal Skateboard* are essentially the same as those used in last night's homework, we recommend that you discuss the activity fully, because the angle analysis is essential for the Ferris wheel problem.

Question 1 of *The Ideal Skateboard* is like a similar question for the Ferris wheel. The skateboarder's speed is approximately 7.33 ft/sec.

Question 2 contains an important element that did not appear in *Homework 27: Racing the River*—namely, the determination of the angle. Students may find it helpful to use a diagram showing an overhead view of the situation, like the one shown here.

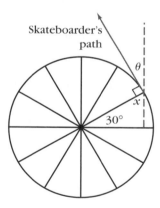

One way to find angle θ is first to find the angle labeled x (from the condition $30° + x = 90°$) and then to solve the equation $x + 90° + \theta = 180°$ (which uses the fact that the tangent is perpendicular to the radius).

Once students see that θ is equal to $30°$, bring out that it's no coincidence that θ is equal to the angle made between the radius to the 3 o'clock position and the radius to the point of release.

Students might then use a diagram like the next one to see that the skateboarder moves 7.33 cos 30° feet, or approximately 6.35 feet, closer to the wall each second. Identify this is as the "toward-the-wall" component of the skateboarder's velocity, similar to the swimmer's toward-shore component in last night's homework.

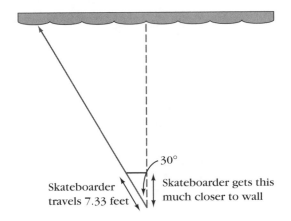

Students will probably find the answer to Question 4 by dividing the distance (30 feet) by the toward-the-wall component of velocity (6.35 feet). This gives a time of approximately 4.7 seconds for the skateboarder to reach the wall.

Question 5 is intended as a confirmation of the component concept. Students should see that the total distance d that the skateboarder travels fits the equation $\cos 30° = \frac{30}{d}$, so $d = \frac{30}{\cos 30°} \approx 34.6$ feet. Dividing this by 7.33 ft/sec again gives about 4.7 seconds.

You can bring out that whichever way students find this time, they use cos 30° as part of the computation. The computation for Question 4 is $\frac{30}{7.33 \cos 30°}$, while the computation for Question 5 is

$$\frac{\left(\frac{30}{\cos 30°}\right)}{7.33}$$

Have students find the "parallel-to-the-wall" component of the skateboarder's velocity. They will probably see that in magnitude, this component is 7.33 sin θ ft/sec. You might use the small triangle in the diagram above to bring out that because the two components are perpendicular, they and the overall speed satisfy the Pythagorean theorem. That is, the values satisfy the equation $(7.33 \sin \theta)^2 + (7.33 \cos \theta)^2 = 7.33^2$.

You might also use this occasion to review the principle that $\sin^2 \theta + \cos^2 \theta = 1$.

- *Components work separately*

It's probably worth mentioning once more the idea that vertical and horizontal components of velocity can be worked with independently of each other. You can remind students of how they worked with these components separately in *High Noon* and *Homework 26: Leap of Faith*.

3. One O'Clock Without Gravity

"How does an angle like θ figure into the Ferris wheel problem?"

Ask how an angle analogous to θ figures into the Ferris wheel problem. Bring out that the release point of the diver determines the angle of his initial motion, so it affects the way his motion breaks up into vertical and horizontal components. (You might bring out that the diver's initial speed is the same no matter when he is released. Similarly, students could find the speed of the skateboarder without first finding θ.)

With this introduction, have students work in groups on *One O'Clock Without Gravity*. The key step relates to the angles in the diagram. Once again, if needed, remind them that the tangent to the circle is perpendicular to the radius.

This activity is scheduled for discussion on Day 29.

Homework 28: General Velocities

Discussion of this assignment will produce the final key formula for the central unit problem.

CLASSWORK

One O'Clock Without Gravity

One night before the premiere of the show, our diver had a dream in which he was merrily spinning around on his Ferris wheel.

In this dream, the assistant decided to let go when the diver was at the 1 o'clock position. But in the diver's dream, there was no gravity, so he sailed up and to the left at a constant speed, in a direction that was tangent to the Ferris wheel's circumference at the 1 o'clock position.

Of course, his speed was the same as his speed when he was on the platform. As you know, that speed is 2.5π feet per second, which is approximately 7.85 feet per second.

1. What was the vertical component of his velocity? That is, how much height did he gain each second? (As in other recent problems, you will need to find the angle labeled θ.)

2. What was the horizontal component of his velocity? (Remember that movement to the left is considered negative.)

General Velocities

As you have seen, although the speed of the Ferris wheel is constant, the vertical and horizontal components of the diver's velocity are different for different positions in the cycle.

1. What are the horizontal and vertical components of the diver's initial velocity if he is released after 8 seconds on the Ferris wheel? (*Reminder:* The period for the Ferris wheel is 40 seconds. Also remember that for horizontal velocity, the positive direction is to the right.)

To generalize the situation, suppose that the diver is released after W seconds.

2. First, assume that W is less than 10, so that the diver is still in the first quadrant when he is released. Write an expression in terms of W for the vertical and horizontal components of the diver's initial velocity.

3. Now consider all values of W from 0 to 40.

   a. For which values of W is the vertical component of velocity positive? For which values is it negative? For which values is it zero?

   b. For which values of W is the horizontal component of velocity positive? For which values is it negative? For which values is it zero?

Release at Any Angle

Mathematical Topics

- Finding general expressions for the vertical and horizontal components of the diver's velocity in the main unit problem
- Studying the motion of falling objects when both the vertical and the horizontal components of initial velocity are nonzero

Outline of the Day

In Class

1. Discuss *One O'Clock Without Gravity* (from Day 28)

- Bring out that the angle between the diver's path and the vertical direction is equal to the angle through which the Ferris wheel turns before the diver is released

2. Discuss *Homework 28: General Velocities*

- Post the first-quadrant formulas for the components of the diver's initial velocity

3. *Release at Any Angle*

- Students examine whether their first-quadrant formulas apply for all angles

4. Discuss *Release at Any Angle*

- Elicit several explanations, including the use of specific examples, for why the first-quadrant formulas apply in general

At Home

Homework 29: A Portfolio of Formulas

Note: We suggest that you begin class with a discussion of yesterday's activity, *One O'Clock Without Gravity,* and then discuss last night's homework.

1. Discussion of *One O'Clock Without Gravity*

You can let different diamond card students share their work on this. The key to the problem is recognizing that the angle between the diver's path and the vertical direction (labeled *y* in the accompanying diagram) is the same as the angle through which the diver has turned, which is the angle between the horizontal radius and the radius to his release point. (That angle of turn is 60° because the diver is released at the 1 o'clock position.) The reasoning is the same as in *The Ideal Skateboard*.

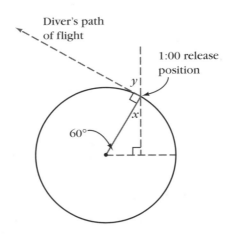

Once students find the angle labeled *y*, the work is similar to what they did in *The Ideal Skateboard*. They might use a diagram like the next one to find the components.

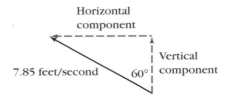

They should see that the vertical component has a magnitude of 7.85 cos 60°, so the diver is rising at about 3.9 ft/sec, and the horizontal component has a magnitude of 7.85 sin 60°, so the diver is moving to the left at about 6.8 ft/sec.

"How do we show that the horizontal travel is to the left or right?"

Be sure to bring out that this horizontal motion is to the left, so that the horizontal component of velocity is −6.8 ft/sec. We suggest that you save the values for these components of velocity for comparison with the results from Question 1 of *Homework 28: General Velocities*.

2. Discussion of *Homework 28:*
General Velocities

Except for the choice of release position, Question 1 of last night's homework is identical to Questions 1 and 2 of *One O'Clock Without Gravity,* so you may want to have only a brief presentation.

The presenter should show that the platform has turned through a 72° angle. Thus, the vertical component of velocity is 7.85 cos 72°, so the diver is initially moving upward at approximately 2.43 ft/sec. The horizontal component of velocity is −7.85 sin 72°, so the diver is initially moving to the left at approximately 7.47 ft/sec.

"Do the differences between the homework results and the results in 'One O'Clock Without Gravity' make sense?"

By comparison, in the discussion of *One O'Clock Without Gravity,* students found the vertical component to be 3.9 ft/sec and the horizontal component to be −6.8 ft/sec. Ask if the differences between the results of the two problems make sense. They should see that the larger angle in the homework problem means that the diver's initial velocity is "more horizontal," and they should recognize that this is consistent with the results.

• *Question 2*

On Question 2, students should be able to generalize their work from Question 1 of the homework and from *One O'Clock Without Gravity* to get the following formulas for the case in which the diver is released within the first quadrant (*W* < 10).

  • Vertical component of velocity = 7.85 cos 9*W*

  • Horizontal component of velocity = −7.85 sin 9*W*

This involves using the expression 9*W* to represent the angle of rotation after traveling for *W* seconds, as done earlier in the unit.

As with the specific examples, be sure students recognize the significance of the minus sign for the horizontal component. They should see that for angles in the first quadrant, the diver's horizontal movement will be to the left, so this component of velocity needs to be negative.

• *Question 3*

For Question 3a, students should see (from a diagram or similar analysis) that the vertical component of the diver's velocity is positive if the diver is released in either the first or fourth quadrant, which means that *W* is either less than 10 or more than 30. Similarly, the vertical component of the diver's velocity is negative if *W* is between 10 and 30, and is zero for both *W* = 10 and *W* = 30.

For Question 3b, students should see that the horizontal component of the diver's velocity is positive if *W* is greater than 20, negative if *W* is less than 20, and zero at 0, 20, and 40.

Note: You may want to post these conclusions, because they will be referred to in the next activity, *Release at Any Angle*.

3. *Release at Any Angle*

In Question 2 of last night's homework, students developed formulas for the components of the diver's initial velocity. But thus far, they have only considered the case in which the diver is released within the first quadrant (that is, for *W* < 10). Tell them that the main goal of *Release at Any Angle* is to examine whether these formulas work for all points of release.

Let students work on this in groups. The goal of Questions 1b and 1c is to verify that a graph based on the formula is consistent with an intuitive, qualitative analysis of the situation. Even if groups have not finished the activity, interrupt to allow a few minutes at the end of class to discuss at least through Question 1c and to assure the class that the formulas do, in fact, work for all values of *W*.

4. Discussion of *Release at Any Angle*

Let a club card student sketch the graph based on the proposed formula for the vertical component of velocity (Question 1a). As a class, you can verify (see Question 1b) that the graph is consistent with the sign analysis from *Homework 28: General Velocities*.

For Question 1c, have another club card student describe, based on the situation itself, how the vertical component of velocity changes in size as *W* increases from 10 to 20. The presenter should note that at *W* = 10, the diver's initial motion is horizontal, so the vertical component is zero, and that as *W* increases, his motion becomes more directly downward so that the size of the vertical component increases (becoming "more negative"). Again, confirm as a class that this is consistent with the graph.

If time allows and you feel that students need the discussion, continue with presentations on Question 1d and then Question 2. The key in Question 1d is to confirm that the vertical component of velocity does, in fact, increase in size as *W* increases from 11 to 12.

For instance, for $W = 11$, the angle of turn is 99° (which is $9W$), and students might use a diagram like the one below.

If they view this diagram sideways so that the upward line is like the positive direction of the x-axis, then the vertical component of the diver's initial motion is like the x-coordinate of a point on the ray showing that motion.

Students have already seen, in discussing the Ferris wheel problem, that they can find a point's x-coordinate by multiplying its r-value by $\cos \theta$. Here, the overall speed, 7.85 ft/sec, is like that r-value, and the vertical component of the diver's initial motion is 7.85 cos 99°. Bring out that this is negative but not very large, just as cos 99° is negative but not very large.

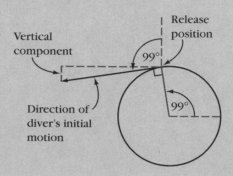

• *Posting the formulas*

Before concluding, assure students that the formulas do work for all angles, and post them along with the other formulas around the room. To have more precise values available for the final solution of the unit problem, we suggest that you use the exact value of 2.5π ft/sec (rather than 7.85 ft/sec) as the speed of the platform as it turns around on the Ferris wheel.

If the diver is released W seconds after passing the 3 o'clock position, then his initial velocity will have these components:

- **Vertical component of velocity = $2.5\pi \cos 9W$**

- **Horizontal component of velocity = $-2.5\pi \sin 9W$**

Homework 29: A Portfolio of Formulas

This assignment is preparation for the solution of the unit problem in class tomorrow. You may want to give students some time to take notes on what's posted around the classroom, in case they haven't taken good notes all throughout the unit.

Point out to students that this summary will be included in their portfolios for this unit.

Release at Any Angle

In *Homework 28: General Velocities*, you examined the components of the diver's initial velocity if he was released within the first quadrant, perhaps using a diagram like the one shown here. This analysis leads to these two equations.

- Vertical component of velocity = 7.85 cos 9*W*

- Horizontal component of velocity = −7.85 sin 9*W*

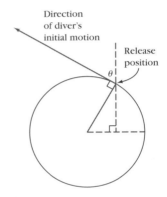

Here, *W* is the time elapsed between when the platform passes the 3 o'clock position and when the diver is released, so the angle of turn is 9*W*.

If the diver is still in the first quadrant, then *W* is less than 10 seconds. But what if *W* is more than 10 seconds? Do these formulas still work?

1. **a.** Assume for the moment that the formula for the vertical component of the diver's initial velocity is correct for all values of *W.* Use that formula to make a graph showing this vertical component of velocity as a function of *W,* from *W* = 0 to *W* = 40.

Continued on next page

b. In Question 3a of *Homework 28: General Velocities*, you examined how the sign of the vertical component of velocity depends on *W*. Is your graph consistent with your results from that question? Explain.

c. For values of *W* between 10 and 20, the diver is released in the second quadrant. Describe how the *size* of the vertical component of velocity changes as *W* increases from 10 to 20. Is your graph consistent with your description? Explain.

d. Find the vertical component of velocity for *W* = 11 and *W* = 12. (Use diagrams as needed to show how to get these values. Do not simply rely on the formula.) Are these values consistent with your conclusions in Question 1c? Explain.

2. Go through a sequence of steps for the horizontal component of velocity similar to Questions 1a through 1d.

A Portfolio of Formulas

To find the correct release time for the circus act, you could take a guess, see what would happen if the diver were released after that much time, and then adjust your guess until you got a "close enough" answer. Or you could develop an equation whose solution would be the answer to the question of when to release the diver.

In preparation for the second approach, your assignment here is to compile a portfolio of all the formulas that have been developed in this unit that might be helpful in solving the unit problem. This portfolio of formulas will be included in your general portfolio for this unit.

Each formula should be appropriately labeled and each variable clearly defined so that you will know what the formula represents.

DAY
30

Diver at Two O'Clock

Mathematical Topics

- Summarizing formulas related to the unit problem
- Analyzing a specific situation related to the unit problem

Outline of the Day

In Class

1. Discuss *Homework 29: A Portfolio of Formulas*

- Have students share formulas, but do not discuss how to use them to solve the unit problem

2. *Moving Diver at Two O'Clock*

- Students do a complete analysis for a Ferris wheel situation involving both vertical and horizontal initial velocity

3. Discuss *Moving Diver at Two O'Clock*

- Use this discussion to review the general principles

At Home

Homework 30: *The Danger of Simplification*

1. Discussion of *Homework 29: A Portfolio of Formulas*

Have volunteers each share one or two of the formulas they compiled in last night's homework, but do not begin a discussion of how to use those formulas. Students will probably use many of them in today's activity, *Moving Diver at Two O'Clock,* and then put them together in tomorrow's activity, *"High Dive" Concluded,* which culminates the unit.

---

2. *Moving Diver at Two O'Clock*

Students can start right in on *Moving Diver at Two O'Clock.* In this activity, they apply the formulas just developed to another Ferris wheel situation. But this time, they follow the details through to find the diver's falling time and both the diver's and the cart's horizontal position when the diver reaches the water level.

Assure students that they can use all the formulas and principles that have been posted around the classroom. (Students sometimes are tempted to develop these formulas again, but it is inefficient for them to ignore all the knowledge that they have accumulated.)

You will probably want fairly detailed presentations on Questions 1 and 2. You may want to give overhead transparencies and pens to groups to use for preparing these presentations.

3. Discussion of *Moving Diver at Two O'Clock*

There are really no new ideas here, but many previously discussed ideas come together in this activity. The following summary is for your benefit. As usual, the ideas should come from the students.

> ● ***Details of the analysis***
> For Question 1, the student will probably go through a sequence of steps like these.
>
> - The diver's height when he is released is $65 + 50 \sin 30° = 90$ feet.
> - The vertical component of his initial velocity is $2.5\pi \cdot \cos 30° \approx 6.80$ ft/sec.
> - The diver's height t seconds after release is given by the formula $90 + 6.80t - 16t^2$.
> - The answer to Question 1 is the positive solution to the equation $90 + 6.80t - 16t^2 = 8$. Students will probably put this equation in standard form as $16t^2 - 6.80t - 82 = 0$ and then solve it using the quadratic formula. The solution is given by the expression
>
> $$\frac{6.80 + \sqrt{6.80^2 + 64 \cdot 82}}{32}$$
>
> which gives $t \approx 2.49$ seconds.
>
> Each part of this analysis represents the application of a formula that has been developed at some point in the unit. You may want to identify the relevant posted formula or result as each part is discussed.

For Question 2, the task is a bit simpler, because the falling time has already been found.

- The diver's x-coordinate when he is released is $50 \cos 30° \approx 43.3$ feet (to the right of center).

- The horizontal component of his initial velocity is $-2.5\pi \cdot \sin 30° \approx -3.93$ ft/sec (that is, he is moving to the left).

- The diver takes 2.49 seconds to fall to the water level.

- The diver moves about $2.49 \cdot (-3.93) \approx -9.8$ feet while falling, so his position when he reaches the water level is about $43.3 - 9.8 = 33.5$, which means he is about 33.5 feet to the right of center.

Again, all of this simply represents the application of ideas previously discussed.

Finally, for Question 3, the analysis will probably go something like this.

- The diver is on the Ferris wheel for $30 \div 9 \approx 3.33$ seconds before being released.

- The diver takes 2.49 seconds to fall to the water level.

- Altogether, the cart is moving for $3.33 + 2.49 = 5.82$ seconds.

- The cart starts 240 feet to the left of center, so its initial position is -240.

- The cart travels 15 ft/sec, so it travels a total of about $5.82 \cdot 15 = 87.3$ feet to the right.

- The cart's position when the diver reaches the water level is $-240 + 87.3 = -152.7$, so the cart is 152.7 feet to the left of center.

Before leaving the problem, be sure to summarize what would actually happen if this scenario actually took place. The diver would land about 33.5 feet to the right of center, while the cart would be 152.7 feet to the left of center, so the cart would miss the diver by 186.2 feet. Let's hope we can avoid such a disaster!

Homework 30: The Danger of Simplification

In this assignment, students examine what will happen to the diver if they use the result from *Moving Cart,* *Turning Ferris Wheel* but take into account the diver's initial velocity due to the motion of the Ferris wheel.

Moving Diver at Two O'Clock

In *Three O'Clock Drop*, you were asked how long it would take for the diver to reach the water level if he were released at the 3 o'clock position. In that problem, the diver's initial velocity had only a vertical component.

In *High Noon*, you answered the same question for a situation in which the initial velocity was all horizontal.

In this activity, you will examine a Ferris wheel situation in which the diver's initial velocity is a blend of vertical and horizontal motion. Specifically, suppose the diver is released after a 30° turn on the Ferris wheel, at the moment when the platform reaches the 2 o'clock position.

1. How long does it take for the diver to reach the water level?

2. What is the diver's x-coordinate when he reaches the water level? (Remember that you are using a horizontal coordinate system in which the base of the Ferris wheel is zero and the positive direction is to the right.)

3. What is the cart's x-coordinate when the diver reaches the water level?

The Danger of Simplification

In *Moving Cart, Turning Ferris Wheel,* you found that the assistant should release the diver after approximately 12.3 seconds. But that analysis was based on the idea that the diver would fall straight down as if released from a stationary platform.

You've now seen that the motion of the Ferris wheel would cause the diver to have an initial speed of 2.5π feet per second (or about 7.85 feet per second), and that this initial speed has both a vertical and a horizontal component. This initial velocity affects both the amount of time the diver is in the air and the diver's x-coordinate at the moment when he reaches the water level.

To learn whether the initial velocity really matters, suppose the diver were released after 12.3 seconds (the time found in *Moving Cart, Turning Ferris Wheel*), and answer these questions, taking the initial velocity into account.

1. How long would it take for the diver to fall to the level of the water in the cart?

2. What would the diver's x-coordinate be at the moment when he reached the water level?

3. What would the cart's x-coordinate be at the moment when the diver reached the water level?

4. Would the diver land in the tub of water?

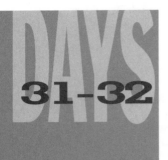

DAYS
31–32

High Dive
Concluded

This page in
the student book
introduces
Days 31 and 32.

*Stephanie Wood, Elizabeth Graf,
Sharon Wang, and
Rochak Nevpane work on
their group's presentation
of the unit's final solution.*

At last! It's time to put all the formulas and ideas
together and figure out when the assistant should let go
of the diver. The diver will certainly appreciate your
hard work and careful analysis.

DAY 31

High Dive Concluded

Students review their solution to the simplified unit problem and begin solving the more complex version.

Mathematical Topics

- Examining the effect of the initial velocity for the value of W that worked for the simplified unit problem
- Beginning the final stage of the solution of the main unit problem

---

Outline of the Day

In Class

1. Discuss *Homework 30: The Danger of Simplification*

- Emphasize that the diver's initial velocity has a significant effect on the result

2. *"High Dive" Concluded*

- Students solve the unit problem
- The activity will be discussed on Day 32

At Home

Homework 31: A Trigonometric Reflection

---

1. Discussion of *Homework 30: The Danger of Simplification*

The main task in this assignment is finding the answer to Question 1. For your reference, we will give the analysis here in some detail. But keep in mind that students will need to generalize this process to solve the unit problem, so be sure that they develop the ideas themselves.

- *Question 1*

To answer Question 1, students need to find the diver's height upon release and the vertical component of his initial velocity. (We also present here his horizontal position upon release and the horizontal component of his initial velocity. Students do not actually need these facts until Question 2, but they are likely to find these values as part of this preliminary stage of the analysis.)

Although students have formulas for these components, this may be the first time that they have used those formulas for a non-first-quadrant case, so you may want to discuss the details. Students should reach these conclusions for $W = 12.3$.

- The vertical component is given by the expression $2.5\pi \cos (9 \cdot 12.3)$, which yields a value of approximately -2.78 ft/sec. (Bring out that this is consistent with the fact that in the second quadrant, the diver is moving downward.)

- The horizontal component is given by the expression $-2.5\pi \sin (9 \cdot 12.3)$, which yields a value of approximately -7.35 ft/sec. (Bring out that this is consistent with the fact that in the second quadrant, the diver is moving to the left.)

You may also want to bring out that the horizontal component is larger in magnitude than the vertical component, which reflects the fact that for $W = 12.3$, the diver is somewhere on the way from the 12 o'clock position to the 11 o'clock position, so his movement is primarily horizontal.

Students also need to determine the diver's position at the moment of release.

- His height above the center of the Ferris wheel is given by the expression $50 \sin (9 \cdot 12.3)$, which yields a value of approximately 46.8 feet. This means that he is $46.8 + 57 = 103.8$ feet above the water level.

- His x-coordinate (relative to the center of the Ferris wheel) is given by the expression $50 \cos (9 \cdot 12.3)$, which yields a value of approximately -17.7. This means that he is about 17.7 feet to the left of center.

Next, students need to apply their work on finding the time for the diver to fall to the water level. They might explicitly write down the expression for his height t seconds after release, which is $103.8 - 2.78t - 16t^2$, and use the quadratic formula to solve the equation obtained by setting this expression equal to 0. Or they might apply the formula for falling time found on Day 26, in which F is expressed by the equation

$$F = \frac{v + \sqrt{v^2 + 64h}}{32}$$

where v is the vertical component of velocity (which is -2.78 ft/sec in this case) and h is the distance the diver falls (which is 103.8 feet in this case).

In either case, students should reach this conclusion for Question 1.

If the diver is released after 12.3 seconds, he will reach the water level in approximately 2.46 seconds.

That is, if $W = 12.3$ seconds, then $F = 2.46$ seconds.

- ## Questions 2 and 3

Once the falling time has been determined, the rest is fairly straightforward. The diver moves to the left at 7.35 ft/sec for 2.46 seconds, starting at $x = -17.7$, so his x-coordinate when he reaches the water level can be found from the expression

$$-17.7 + 2.46 \cdot (-7.35)$$

This comes out to approximately -35.8. That is, the diver reaches the water level approximately 35.8 feet to the left of the center of the Ferris wheel.

For Question 3, the cart moves to the right at 15 ft/sec both while the diver is on the Ferris wheel (12.3 seconds) and while he is falling (2.46 seconds). The cart starts at $x = -240$, so its x-coordinate at the time the diver reaches the water level is found from the expression

$$-240 + (12.3 + 2.46) \cdot 15$$

This comes out to approximately -18.6, so at the moment the diver reaches the water level, the cart is approximately 18.6 feet to the left of the center of the Ferris wheel.

- ## Question 4

You may want to play up the drama of the conclusion from Question 4. Students should have seen that the diver misses the cart by approximately 17.2 feet! This should demonstrate the significance of the work students have been doing in considering the diver's initial velocity.

Of course, we have never specified the dimensions of the tub of water. Perhaps the tub is wide enough to deal with this level of miscalculation.

- ## Looking back

You may want to compare these results with those from *Moving Cart, Turning Ferris Wheel* (discussed on Day 16). In that simplified version of the problem, the diver's x-coordinate throughout his fall was -17.5, and his falling time was about 2.55 seconds.

Bring out that the vertical component of the diver's initial velocity shortened his falling time slightly (from 2.55 seconds to 2.46 seconds), which means that the cart is not as far along the track. (It's at -18.6 instead of -17.5.) But the horizontal component of the diver's initial velocity had a substantial effect. He ends up at $x = -35.8$, instead of at $x = -17.5$, more than 18 feet farther to the left. Ouch!

2. *"High Dive" Concluded*

Toot a trumpet! Here we go!

Homework 31: A Trigonometric Reflection

Tell students that this summary will be included in their portfolios for this unit, along with the summary they did in *Homework 29: A Portfolio of Formulas*.

"The students struggled at the end of the unit but wound up wowing me and themselves when they put together all of the dive's components. It felt great!"

IMP teacher Moe Burkhart

Drew Gilliam ponders how to pull together all the information he's developed and the different formulas he's collected.

High Dive
Concluded

It's now time, once again, to solve the main unit problem. Although you solved a simplified version in *Moving Cart, Turning Ferris Wheel,* now you will take into account that the diver leaves the Ferris wheel with some initial velocity, with both horizontal and vertical components.

Again, here are the details about the setup for the act.

• The Ferris wheel has a radius of 50 feet.

• The center of the Ferris wheel is 65 feet above the ground.

• The Ferris wheel turns counterclockwise at a constant rate, making a complete turn every 40 seconds.

• When the cart starts moving, it is 240 feet to the left of the base of the Ferris wheel.

• The cart moves to the right along the track at a constant speed of 15 feet per second.

• The water level in the cart is 8 feet above the ground.

• The cart starts moving as the platform passes the 3 o'clock position.

As before, let $t = 0$ represent the time when the Ferris wheel passes the 3 o'clock position, which is also when the cart begins moving. Let W represent the number of seconds that elapse between $t = 0$ and the moment when the diver is released. Your assignment is to answer this question.

> *For what choice of W will the diver land in the tub of water?*

Continued on next page

Feel free to use all the accumulated formulas and shortcuts that you have developed over the course of this unit.

Your group should prepare an oral report on your conclusions and how you reached them. You should also prepare your own write-up of your solution.

Suggestion: You may want to give variable names to complex expressions that are part of your solution. If you use a calculator to graph an equation, you can define some of these expressions as preliminary functions and then express your main equation or function in terms of these.

A Trigonometric Reflection

While you're pondering the final details of the solution to the unit problem, take time to think about what you have learned about trigonometry in this unit.

In addition to the general definitions of the sine and cosine functions, you have also learned some things about graphs, trigonometric identities, and polar coordinates.

Compile a summary of these ideas. You should include diagrams as needed to explain any formulas. You do not need to include formulas that relate only to the unit problem, but you should use ideas about the Ferris wheel to explain ideas about trigonometry.

DAY 32

Students complete the unit problem.

High Dive Concluded

Mathematical Topics

- Concluding the main unit problem

Outline of the Day

In Class

1. Discuss *Homework 31: A Trigonometric Reflection*

2. Discuss *"High Dive" Concluded* (from Day 31)

- Have presentations on the unit problem

At Home

Homework 32: "High Dive" Portfolio

1. Discussion of *Homework 31: A Trigonometric Reflection*

You may want to focus on the process students used in extending the trigonometric functions from the right-triangle definitions to the complete functions defined in this unit.

2. Discussion of *"High Dive" Concluded*

Let groups finish up their work and write-ups on this activity. You may want to give hints to groups as they proceed, including how to define some functions in terms of others to avoid horrendously complex expressions on their calculators.

You may wish to have different groups do various parts of the presentation so that it doesn't get tedious. You might have a group that did not finish the problem present some of what it did first, and then have other groups follow up to describe how they continued from that point.

If students did not get the correct answer, we suggest that you let them rework their report for homework. Students may want to use their reports in studying for the assessment or for preparing their portfolios. The reports should be part of everyone's portfolio.

• An outline of the solution

Here is an outline of a possible approach to solving the problem. It is a synthesis of ideas developed over the course of the unit.

As indicated in the activity itself, $t = 0$ represents the time when the cart begins moving and $t = W$ represents the time at which the diver is released. The variables described here represent other components of the solution.

- h = the diver's height above the water level at the time of release

$$h = 57 + 50 \sin 9W$$

- c = the diver's x-coordinate at the time of release

$$c = 50 \cos 9W$$

- v_y = the vertical component of the diver's velocity when he is released

$$v_y = 2.5\pi \cos 9W$$

- v_x = the horizontal component of the diver's velocity when he is released

$$v_x = -2.5\pi \sin 9W$$

- F = the duration of the diver's fall from the time of release until he reaches the water level

$$F = \frac{v_y + \sqrt{v_y{}^2 + 64h}}{32}$$

Reminder: This expression comes from using the expression $h + v_y t - 16t^2$ for the diver's height above the water level t seconds after he is released, and solving the equation $h + v_y t - 16t^2 = 0$. In standard form, this equation is $16t^2 - v_y t - h = 0$, so for the purposes of the quadratic formula, we have $a = 16$, $b = -v_y$, and $c = -h$. Students will have to recall that in applying the quadratic formula in this context, they want the "+" portion of the $\pm$ sign.

Based on these variables, students can find expressions for other aspects of the problem.

- $W + F$ = the total time the cart is moving

- $-240 + 15(W + F)$ = the x-coordinate of the cart at the time the diver reaches the water level

- $c + v_x F$ = the x-coordinate of the diver when he reaches the water level

The task is to find the value of *W* that puts the cart in the right place at the right time. Based on the two expressions just given for the position of the cart and the diver at the time the diver reaches the water level, we need to find the value of *W* that solves the equation

$$-240 + 15(W + F) = c + v_xF$$

It turns out that the desired value is $W \approx 11.45$ seconds. If the diver is released 11.45 seconds after the cart starts, then he lands in the water about 30.53 feet to the left of the base of the Ferris wheel.

You may want to ask students to trace this value of *W* through the problem.

For your convenience, here are the values of the different pieces of the puzzle. (*Note:* These results were found using a more precise value for *W* of 11.449 seconds. Final values were rounded to the nearest hundreth, but values for expressions such as $W + F$, $c + v_xF$, and $-240 + 15(W + F)$ were initially found using pre-round-off values for the variables used in those expressions.)

- $h = 105.71$ (the diver is 105.71 feet above the water level when released)

- $c = -11.28$ (the diver is 11.28 feet to the left of center when released)

- $v_y = -1.77$ (the diver has an initial vertical component of velocity of 1.77 ft/sec downward)

- $v_x = -7.65$ (the diver has an initial horizontal component of velocity of 7.65 ft/sec to the left)

- $F = 2.52$ (the diver is in the air for 2.52 seconds)

- $W + F = 13.96$ (the cart travels for 13.96 seconds)

- $c + v_xF = -30.53$ (the diver is 30.53 feet to the left of center when he reaches the water level)

- $-240 + 15(W + F) = -30.53$ (the cart is 30.53 feet to the left of center when the diver reaches the water level)

Hurrah! It works!

You may want to have students compare this result with the analysis they did in both *Moving Cart, Turning Ferris Wheel* and *Homework 30: The Danger of Simplification.*

- *A whole new problem*

    You might want to point out to students that we have ignored a whole other problem, which is more a question of physiology than mathematics or physics:

    Could the diver survive the dive into the water?

Homework 32: "High Dive" Portfolio

For homework tonight, students will complete their unit portfolios. They will have done part of the selection process in last night's homework, so their main task in this assignment is to write their cover letters.

Be sure that students bring in their portfolios tomorrow with their cover letters as the first item. They should also bring to class any other work that they think will be of help on tomorrow's unit assessments. The remainder of their work can be kept at home.

High Dive Portfolio

Now that *High Dive* is completed, it is time to put together your portfolio for the unit. Compiling this portfolio has three parts.

- Writing a cover letter in which you summarize the unit

- Choosing papers to include from your work in this unit

- Discussing your personal mathematical growth in this unit

Cover Letter for *High Dive*

Look back over *High Dive* and describe the central problem of the unit and the main mathematical ideas. This description should give an overview of how the key ideas, such as extending the sine and cosine functions and finding falling-time functions, were developed and how they were used to solve the central problem.

In compiling your portfolio, you will select some activities that you think were important in developing the key ideas of this unit. Your cover letter should include an explanation of why you selected the particular items.

Continued on next page

Selecting Papers from *High Dive*

Your portfolio for *High Dive* should contain these items:

- *"High Dive" Concluded*

- A Problem of the Week

  Select one of the POWs you completed in this unit (*Tower of Hanoi* or *Paving Patterns*)

- *Homework 29: A Portfolio of Formulas*

- *Homework 31: A Trigonometric Reflection*

- Other key activities

  Identify two concepts that you think were important in this unit. For each concept, choose one or two activities that helped improve your understanding, and explain how the activity helped.

Personal Growth

Your cover letter for *High Dive* describes how the mathematical ideas develop in the unit. As part of your portfolio, write about your own personal development during this unit. You may want to address this question.

> *How do you feel about your ability to solve a problem that is as complex and that has as many components as the "High Dive" problem?*

You should include here any other thoughts about your experiences with this unit that you want to share with a reader of your portfolio.

DAY 33

Final Assessments

Students do the in-class assessment and begin work on the take-home assessment.

Outline of the Day

In Class

Introduce assessments

• Students do *In-Class Assessment for "High Dive"*
• Students begin T*ake-Home Assessment for "High Dive"*

At Home

Students complete *Take-Home Assessment for "High Dive"*

End-of-Unit Assessments

Note: The in-class assessment is intentionally quite short so that time pressures will not be a factor in students' ability to do well. The IMP *Teaching Handbook* contains general information about the purpose of the end-of-unit assessments and how to use them.

Tell students that today they will get two tests—one that they will finish in class and one that they can start in class and will be able to finish at home. The take-home part should be handed in tomorrow.

Tell students that they are allowed to use graphing calculators, notes from previous work, and so forth, when they do the assessments. (They will have to do without graphing calculators on the take-home portion unless they have their own.)

The assessments are provided separately in Appendix B for you to duplicate.

In-Class Assessment for *High Dive*

Walter the whale is a mathematical sort of creature. He swims with a periodic rise and fall, patterning his swimming path after a sine curve.

On a particular day, Walter is swimming with an amplitude of 15 feet. More specifically, when he is at his highest point, his back rises 5 feet out of the water. Then he dives down to a point where his back is 25 feet below the surface level.

Of course, he then comes back up again, and then goes down again, and so on. It takes about 20 seconds from the time Walter hits his high point until he reaches his lowest point.

You come out on the deck of a ship and look out to sea exactly in the direction where Walter is swimming. What is the probability that you will see him the instant you look?

Take-Home Assessment for *High Dive*

1. Do you remember Larry, the stunt diver from the Year 3 unit *Small World, Isn't It?* His most famous dive is off a cliff into the ocean near the town of Poco Loco. Larry dives from a 60-foot cliff. He begins with a magnificent jump that gives him an initial vertical component of velocity of 6 feet per second (upward).

   Figure out how long Larry's dive will take, giving your answer in two ways.

   • To the nearest hundredth of a second

   • As an exact value, using a square root if necessary

   (Assume that once Larry leaves the cliff, his vertical motion is affected only by the force of gravity.)

2. In this unit, you've seen that the sine, cosine, and tangent functions can be extended to arbitrary angles using the coordinate system and a diagram like the one at the right. Explain how to use the diagram to define the *secant* function for all angles.

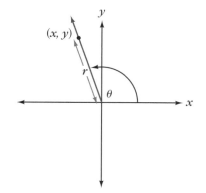

   Reminder: In a right triangle like the one shown here, the secant function is defined by the equation $\sec \theta = \frac{AB}{AC}$.

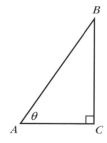

DAY
34

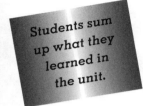

Students sum up what they learned in the unit.

Summing Up

Mathematical Topics

- Summarizing the unit

Outline of the Day

1. Discuss unit assessments

2. Sum up the unit

Note: The assessment discussions and unit summary are presented as if you will use them on the day following the assessments, but you may prefer to wait until after you have looked over students' work on the assessments. These discussion ideas are included here to remind you that you should allot some time for this type of discussion.

1. Discussion of Unit Assessments

Ask for volunteers to explain their work on each of the problems. Encourage questions and alternate explanations from other students.

- *In-class assessment*

  You might have the presenter begin by sketching a graph of the height of Walter's back. The choice of what point in the cycle to use for $t = 0$ is arbitrary, and students' graphs will vary depending on this choice. Using $t = 0$ to represent a time when Walter is at the midpoint in his cycle, on his way up (when his back is 10 feet below the surface), the graph of one complete cycle would look like the one shown here.

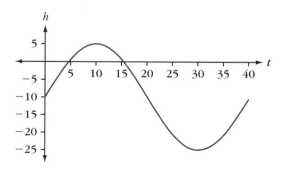

In terms of this graph, students need to know what fraction of the curve is above the horizontal axis. They might do this by solving the equation $15 \sin \theta = 10$, which gives the approximate solutions $\theta = 42°$ and $\theta = 138°$. The difference, $96°$, is roughly 27 percent of $360°$, so the probability that a person will see Walter immediately is approximately .27.

- *Take-home assessment*

    Question 1 is a straightforward application of the ideas of the unit, in which the initial height is 60 feet and the initial (vertical) velocity is 6 ft/sec. Thus, Larry's height after t seconds is $60 + 6t - 16t^2$, and students need to find the positive value of t that makes this expression equal to 0.

    This is equivalent to solving the quadratic equation $16t^2 - 6t - 60 = 0$, and the quadratic formula gives the exact value of the positive solution as

    $$\frac{6 + \sqrt{(-6)^2 - 4 \cdot 16 \cdot (-60)}}{32}$$

    This simplifies to $\frac{6 + \sqrt{3876}}{32}$, which, to the nearest hundredth, is 2.13 seconds.

    On Question 2, students should see that for first-quadrant angles, $\sec \theta$ is equal to the ratio $\frac{r}{x}$, and conclude that this ratio should be used for all angles. They may bring out that in right triangles, $\sec \theta$ is equal to $\frac{1}{\cos \theta}$, and use that relationship as an additional justification for the general definition of the secant function. (If students do not bring out that $\sec \theta$ is equal to $\frac{1}{\cos \theta}$, you should mention this yourself.)

    You may want to take a minute to discuss the graph of the secant function. Bring out that the function is undefined for $\theta = 90°$ and $270°$ and that these are the values for which the cosine function is 0. The accompanying diagram shows how the graph should look.

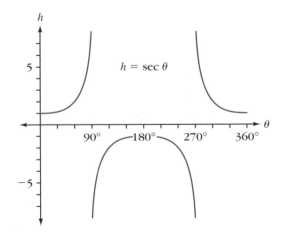

- *Extending cotangent and cosecant*

"Which trigonometric functions haven't you extended yet?"

To complete the work of extending the trigonometric functions, ask students to name the trigonometric functions that they have not yet extended. Review the fact that there are six ratios possible using the lengths of the sides of a right triangle (perhaps reviewing the notation $_3P_2$ at the same time).

Go over the right-triangle definitions for the other two functions, cotangent and cosecant, and the fact that each is the reciprocal of one of the "basic" functions, tangent and sine. Also take a few minutes to look at the graph of each of these functions, bringing out that the graph of the cotangent function is a shifted and reversed version of the graph of the tangent function, and that the graph of the cosecant function is a shifted version of the graph of the secant function.

2. Unit Summary

Let volunteers share their portfolio cover letters as a way to start a discussion to summarize the unit.

Supplemental Problems

This appendix contains a variety of activities that you can use to supplement the regular unit material. These activities fall roughly into two categories.

- Reinforcements, which are intended to increase students' understanding of and comfort with concepts, techniques, and methods that are discussed in class and that are central to the unit

- Extensions, which allow students to explore ideas beyond the basic unit and which sometimes deal with generalizations or abstractions of ideas that are part of the main unit

The supplemental activities are given here in the approximate sequence in which you might use them in the unit. In the student book, they are placed in the same order following the regular materials for the unit.

Here are some specific recommendations about how each activity might work within the unit. (For more ideas about the use of supplemental activities in the IMP curriculum, see the IMP *Teaching Handbook*.)

Mr. Ferris and His Wheel (extension)

In this problem, students can do some background research about the Ferris wheel and any related ideas that interest them. This activity can be assigned at the beginning of the unit.

A Shifted Ferris Wheel (extension)

This activity explores the impact of changing the "starting time" for the Ferris wheel on the function describing the platform's height. It can be assigned after Day 5, once students have seen the general definition of the sine function and its relationship to height on the Ferris wheel.

Prisoner Revisited (reinforcement)

This activity is somewhat similar to *Homework 6: Sand Castles* and *Homework 7: More Beach Adventures* and can be assigned on Day 8.

Lightning at the Beach on Jupiter (reinforcement)

You can use this activity prior to (or after) Day 8 as a review of the basic relationships among rate, time, and distance.

The Derivative of Position (extension)

In *Free Fall,* students use the method of "averaging the endpoints" to develop a formula for the height of a falling object. In this follow-up, they are asked to use derivatives to confirm that formula by showing that an object whose height fits this formula has the correct rate of acceleration.

Polar Equations (extension)

On Day 17, students get a brief introduction to polar coordinates. This problem begins with a review of the definition and some basic exercises, and then introduces them to the graphing of polar equations.

Circular Sine (extension)

In this problem, students will need to do some algebra to transform a polar equation into its rectangular equivalent. (Students should have worked on the supplemental problem *Polar Equations* before doing this problem.)

A Polar Exploration (extension)

This is an open-ended follow-up to the previous two supplemental problems.

A Shift in Sine (extension)

In this activity, students are asked to formalize and prove the observation that the graph of the sine function is a "shift" of the graph of the cosine function.

More Pythagorean Trigonometry (extension)

This activity, a follow-up to *Pythagorean Trigonometry* (Day 18), asks students to develop other Pythagorean identities.

Derivative Components (extension)

Based on a series of activities, students develop formulas on Day 29 for the vertical and horizontal components of the diver's initial velocity. In this activity, they use derivatives to confirm these formulas.

Appendix

Supplemental Problems

This page in the student book introduces the supplemental problems.

The supplemental problems in *High Dive* focus primarily on the trigonometric functions and their relationship with the Ferris wheel situation. Other activities look at the use of derivatives to understand issues in this unit. Here are some examples.

• *A Shifted Ferris Wheel* examines how changing the "starting time" for the Ferris wheel would affect the function describing the platform's height.

• *Polar Equations* and *Circular Sine* continue the work with polar coordinates.

• *The Derivative of Position* uses derivatives to get insight into the formula for the height of a falling object, while *Derivative Components* uses derivatives to understand the vertical and horizontal components of the diver's initial velocity.

SUPPLEMENTAL
PROBLEM

Mr. Ferris and His Wheel

The Ferris wheel is named after its inventor, George Washington Gale Ferris.

Most of us have probably either ridden on or watched a Ferris wheel at some point in our lives. But even though the Ferris wheel turns within our own memories, it's unlikely that we have looked into its fascinating history. Here are some questions about the Ferris wheel that you may want to research. Who was George Ferris? Where was he raised and educated? How did he come to invent the Ferris wheel? What was the first Ferris wheel made of? How did the invention of the Ferris wheel change Ferris's life? Did Ferris make any other notable inventions?

You may want to expand this topic to a broader study of amusement park rides or some other aspect of carnivals and fairs.

SUPPLEMENTAL
PROBLEM

A Shifted Ferris Wheel

In the main unit problem, the diver's platform is at the 3 o'clock position when the cart starts moving. Based on using this moment as $t = 0$, the platform's height after t seconds is given by the expression $65 + 50 \sin 9t$.

Suppose instead that at $t = 0$, the platform was at the 6 o'clock position.

1. Find an expression that would give the platform's height as a function of t.

2. Sketch the graph of the height function, and compare it to the graph for the main unit problem.

3. Consider other positions for the Ferris wheel at $t = 0$, and describe in general how changing the position affects the function describing the platform's height.

SUPPLEMENTAL PROBLEM

Prisoner Revisited

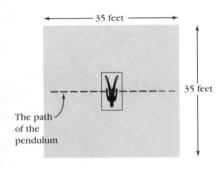

35 feet

35 feet

The path of the pendulum

Do you remember the prisoner from the Year 1 unit *The Pit and the Pendulum*? Well, he's back.

This time, he is lying on the table in the middle of a square prison cell that is 35 feet by 35 feet, again with a pendulum moving back and forth above him, as shown here. (There are no rats in this cell.) As before, the pendulum has a blade at its end. This time, the length of the pendulum does not change, but the table is gradually rising, moving the prisoner up toward the blade.

As he lies there, the prisoner notices that the pendulum's motion is following a sine-like pattern as it swings back and forth. Specifically, the pendulum's horizontal distance $p(t)$ from the center of the cell (measured in feet) is given by the function

$$p(t) = 15 \sin 60t$$

where t is the number of seconds the pendulum has been swinging.

The accompanying diagram shows this horizontal distance, viewed from the front of the cell. (This diagram is not drawn to scale.)

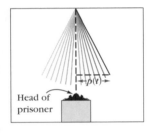

Head of prisoner

$p(t)$

Continued on next page

100 *High Dive*

Interactive Mathematics Program

Suddenly, a friend of the prisoner appears in the adjacent cell. The friend rushes over to the bars that separate the two cells, so that the pendulum swings alternately toward him and away from him, as shown below. (This is an overhead view.)

The prisoner's friend realizes that if the pendulum comes within 3 feet of the bars between the cells, he can reach through the bars, grab onto the pendulum, and stop its motion.

1. Will the pendulum come close enough to the bars between the cells so that the prisoner's friend can reach it? Explain your answer.

2. If so, for how long will the pendulum be in the friend's range each time it comes by?

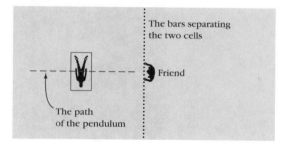

SUPPLEMENTAL
PROBLEM

Lightning at the Beach on Jupiter

As you may have noticed, the three variables of *rate, time,* and *distance* are closely related to one another.

In the problems in this assignment, you are given information about two of these variables. Your task is to find out the value for the third.

Continued on next page

1. Light travels at about 186,000 miles per second. Jupiter is about 483,000,000 miles from the sun. (It's sometimes closer and sometimes farther away, but we'll use this average distance.)

   How long does it take for light to get from the sun to Jupiter?

2. Amparo wants to spend the day at the beach, which is 100 miles away. She leaves at 8:00 in the morning and needs to be home by 7:00 that evening.

   If she wants to have 6 hours at the beach, what should be her average speed for the trip?

3. You see a flash of lightning. About 6 seconds later, you hear the crash of thunder. Assume that the light reaches you instantly and that the sound travels at about 1100 feet per second. (The exact speed of sound depends on atmospheric conditions like temperature.)

   How far away was the lightning?

4. What general relationships exist among rate, distance, and time?

5. How do the concepts of rate, distance, and time relate to the main unit problem?

SUPPLEMENTAL
PROBLEM

The Derivative of Position

In *Free Fall*, you developed an important general principle about free-falling objects. Here is the principle.

If an object falls freely from rest from a height of *h* feet, then its height after *t* seconds is approximately $h - 16t^2$ feet.

This principle builds on the fact from physics that a free-falling object accelerates at approximately 32 feet per second for each second it falls.

Your task in this activity is to confirm the formula $h - 16t^2$ using derivatives, by showing that if an object's height fits this formula, then its acceleration will actually be 32 feet per second for each second it falls.

1. Suppose, as just described, that a certain object is moving downward so that its height $f(t)$ after t seconds is given by the equation $f(t) = h - 16t^2$.

   a. At what rate is the object's height changing at $t = 1$? That is, what is the object's instantaneous velocity at $t = 1$?

   b. Explain why the number you found in Question 1a is the same as the derivative of f at $t = 1$.

2. **a.** Find the derivative of f at $t = 2$, at $t = 5$, and at $t = 10$.

   b. Based on your answers to Questions 1a and 2a, give a general expression for $f'(t)$ in terms of t.

3. What does your result from Question 2b say about the object's acceleration?

Polar Equations

Over the years, you've worked with many equations involving x and y and found the graphs of those equations. The graph of such an equation consists of all those points whose rectangular coordinates fit the equation.

For example, the point with rectangular coordinates $(3, 4)$ is on the graph of the equation $5x - 2y = 7$ because $5 \cdot 3 - 2 \cdot 4 = 7$.

You can also find the graphs for equations involving polar equations. As with equations using x and y, the graph of an equation involving r and θ consists of those points whose polar coordinates fit the given equation.

For example, the point with polar coordinates $(3, 90°)$ is on the graph of the equation $r + \sin \theta = 4$ because $3 + \sin 90° = 4$.

For each of these equations, first find some number pairs for r and θ that fit the equation. Then use those number pairs as polar coordinates and plot the points they represent. Finally, use these points to sketch a graph of the equation. Find more solutions if you need them in order to get a good idea of what the graph should look like.

1. $r = \theta$

2. $r = \cos \theta$

3. $r = 2$ (*Hint:* Where's θ? Think about equations like $x = 3$.)

4. $\theta = 20°$ (*Hint:* See the hint for Question 3.)

SUPPLEMENTAL PROBLEM

Circular Sine

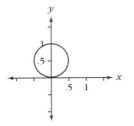

If you were to plot some points for the polar coordinate equation $r = \sin \theta$ and then connect them, you might find that the graph looked something like the diagram shown here. It appears to be a circle, but it's hard to tell for sure merely from plotting a bunch of points and connecting them with a freehand sketch.

Your challenge in this problem is to show that the graph of the polar equation $r = \sin \theta$ is definitely a circle.

1. If this graph is a circle, what are the rectangular coordinates of its center and what is its radius?

2. What is the rectangular equation for the circle whose center and radius you found in Question 1? (*Hint:* Suppose a point (x, y) is on this circle. Use the Pythagorean theorem to get an equation for the distance from this point to the center of the circle.)

3. How can you use the relationships between rectangular and polar coordinates to see that the rectangular equation for Question 2 is equivalent to the polar equation $r = \sin \theta$?

A Polar Exploration

If you worked on the supplemental problem *Polar Equations*, you may have seen that the graphs of simple polar equations, such as $r = \theta$, can give very different graphs from simple equations with rectangular coordinates.

The diagram below shows one of the interesting graphs that you can get from a fairly simple polar equation.

Your assignment on this problem is to investigate graphs and equations using polar coordinates.

Feel free to consult a trigonometry textbook for ideas of interesting equations to explore. Your report on this problem should indicate any references you used and show clearly which ideas came from other books and which were your own.

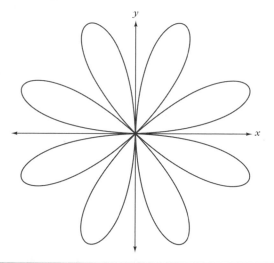

SUPPLEMENTAL PROBLEM

A Shift in Sine

You have observed that the graphs of the functions $z = \sin t$ and $z = \cos t$ are quite similar. One way to describe the relationship is that if you "shift" the graph of the sine function 90° to the left, you get the graph of the cosine function.

1. Express this relationship between the graphs as a trigonometric identity, writing $\sin \theta$ as the cosine of a different angle.

2. Prove the identity you found in Question 1. (*Suggestion:* Use the relationship $\sin \theta = \cos (90° - \theta)$.)

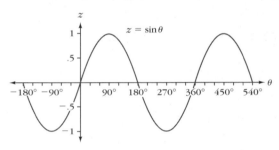

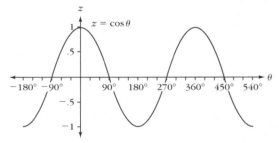

More Pythagorean Trigonometry

The Pythagorean theorem states that in a right triangle such as the one shown here, the lengths of the sides satisfy the equation $a^2 + b^2 = c^2$.

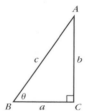

The sine and cosine functions are defined for right triangles by the equations $\sin \theta = \frac{b}{c}$ and $\cos \theta = \frac{a}{c}$.

In *Pythagorean Trigonometry*, you developed an identity involving the sine and cosine functions that resembles the statement of the Pythagorean theorem.

1. State that identity and explain it for acute angles based on the equations $a^2 + b^2 = c^2$, $\sin \theta = \frac{b}{c}$, and $\cos \theta = \frac{a}{c}$.

2. Develop similar identities involving the other trigonometric functions—tangent, cotangent, secant, and cosecant—based on the definitions $\tan \theta = \frac{b}{a}$, $\cot \theta = \frac{a}{b}$, $\sec \theta = \frac{c}{a}$, and $\csc \theta = \frac{c}{b}$.

SUPPLEMENTAL
PROBLEM

Derivative Components

In our standard Ferris wheel, the radius is 50 feet, the Ferris wheel turns counterclockwise at a constant rate with a period of 40 seconds, the center of the Ferris wheel is 65 feet off the ground, and the platform is at the 3 o'clock position at $t = 0$.

Based on these facts, you developed the formula that the platform's height off the ground after t seconds is given by the equation $h(t) = 65 + 50 \sin 9t$.

Of course, this platform is moving, so its height is constantly changing. You found that the vertical component of the platform's velocity is given by the expression $2.5\pi \cos 9t$. Your task in this activity is to confirm this relationship using derivatives.

1. **a.** Pick a specific value for t.

   b. Find $h(t)$ for that value of t.

   c. Find the rate at which the height is changing by finding $h'(t)$ for that value of t.

2. Compare your answer to Question 1c with the value of the expression $2.5\pi \cos 9t$.

3. Repeat Questions 1 and 2 for a different value of t.

4. Explain why the observations you made in Questions 2 and 3 should hold true.

5. Develop a similar sequence of steps to confirm that the horizontal component of the platform's velocity is $-2.5\pi \sin 9t$.

Blackline Masters

This appendix contains these materials for the unit.

- A diagram for explaining the "clock" labels used initially in *Homework 1: The Ferris Wheel*

- A diagram illustrating the setup of the Ferris wheel act, for use from Day 2 on

- A blank coordinate system for use in the Day 5 discussion of *Homework 4: Graphing the Ferris Wheel*

- A completed version of the graph for *Homework 4: Graphing the Ferris Wheel* (for Day 5)

- A graph of the function $z = \sin \theta$ (for Day 6)

- A diagram for use in the discussion of Question 2 of *Distance with Changing Speed* (for Day 8)

- A blank coordinate system for use in the Day 12 discussion of *Homework 11: Where Does He Land?*

- A diagram for use in discussing the motion of an object released from a circular path (for Day 21)

- A diagram for use in the Day 28 discussion of *Homework 27: Racing the River*

- The in-class and take-home assessments for the unit

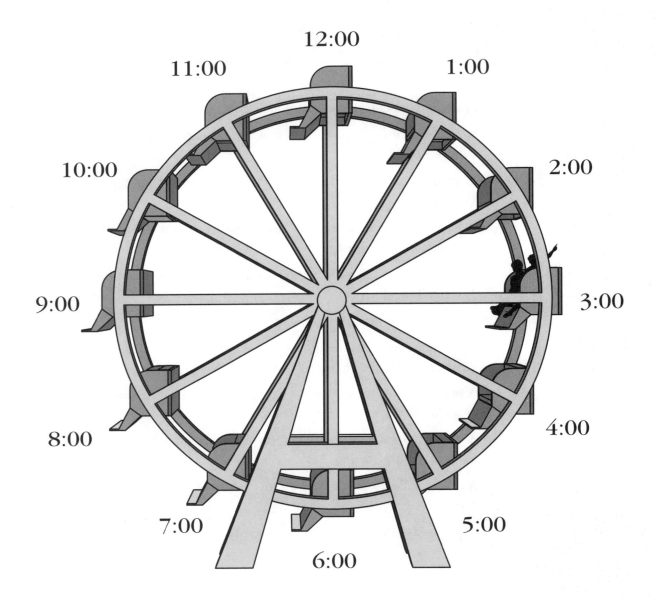

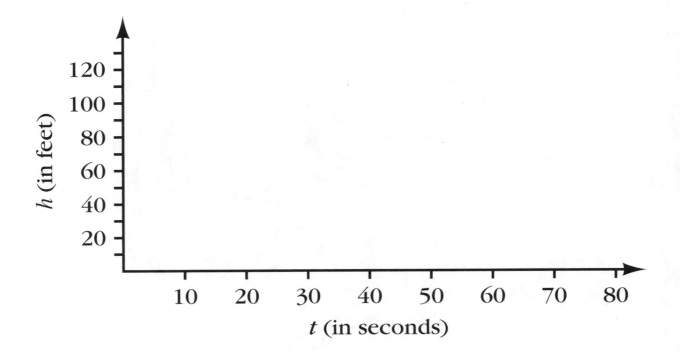

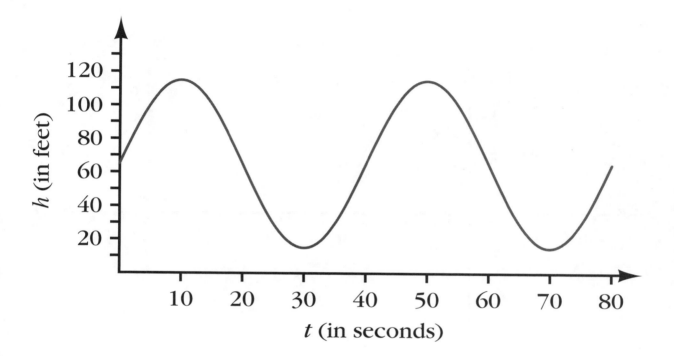

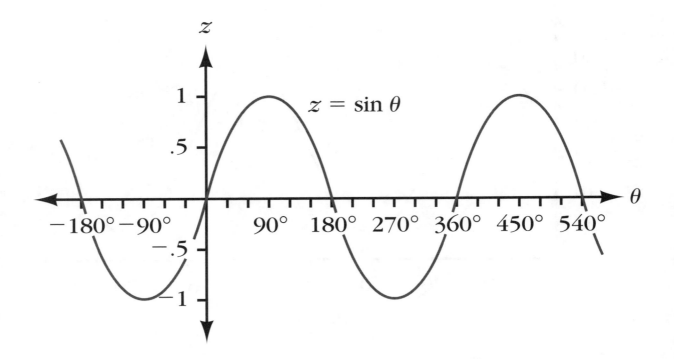

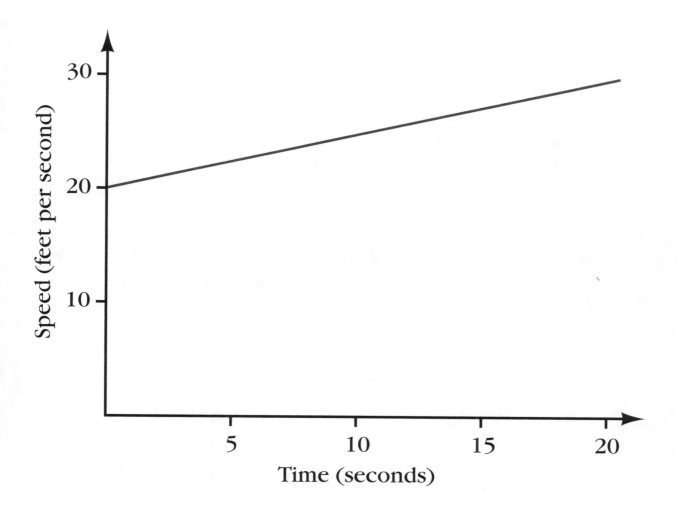

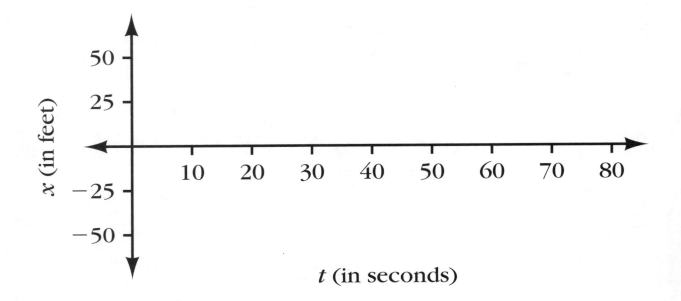

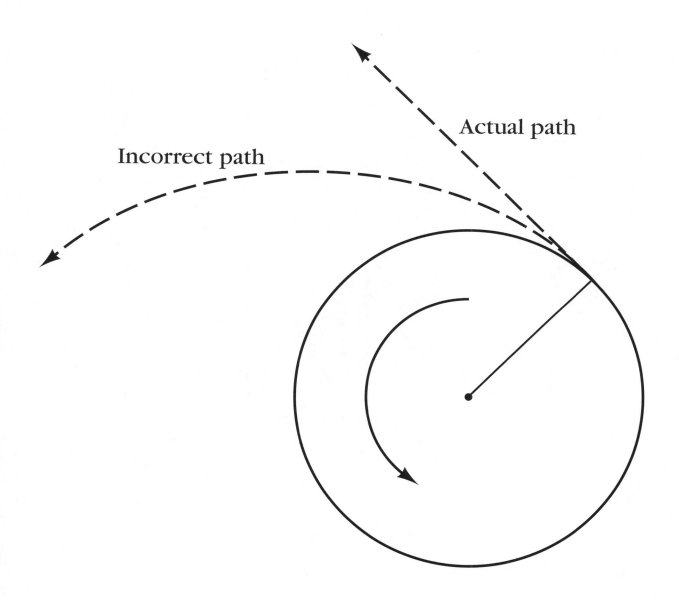

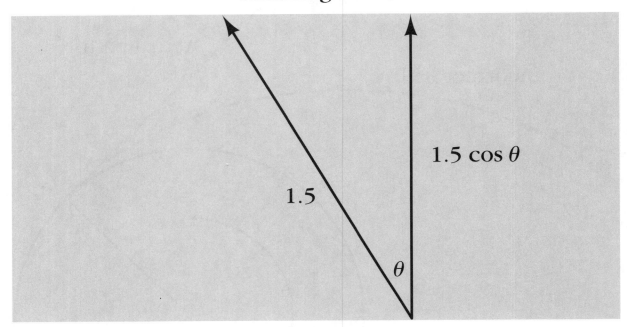

In-Class Assessment
for *High Dive*

Walter the whale is a mathematical sort of creature. He swims with a periodic rise and fall, patterning his swimming path after a sine curve.

On a particular day, Walter is swimming with an amplitude of 15 feet. More specifically, when he is at his highest point, his back rises 5 feet out of the water. Then he dives down to a point where his back is 25 feet below the surface level.

Of course, he then comes back up again, and then goes down again, and so on. It takes about 20 seconds from the time Walter hits his high point until he reaches his lowest point.

You come out on the deck of a ship and look out to sea exactly in the direction where Walter is swimming. What is the probability that you will see him the instant you look?

Take-Home Assessment for *High Dive*

1. Do you remember Larry, the stunt diver from the Year 3 unit *Small World, Isn't It?* His most famous dive is off a cliff into the ocean near the town of Poco Loco. Larry dives from a 60-foot cliff. He begins with a magnificent jump that gives him an initial vertical component of velocity of 6 feet per second (upward) .

Figure out how long Larry's dive will take, giving your answer in two ways.

• To the nearest hundredth of a second

• As an exact value, using a square root if necessary

(Assume that once Larry leaves the cliff, his vertical motion is affected only by the force of gravity.)

2. In this unit, you've seen that the sine, cosine, and tangent functions can be extended to arbitrary angles using the coordinate system and a diagram like the one at the right. Explain how to use the diagram to define the *secant* function for all angles.

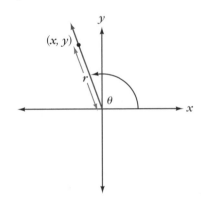

Reminder: In a right triangle like the one shown here, the secant function is defined by the equation $\sec \theta = \frac{AB}{AC}$.

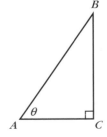

© 2000 Interactive Mathematics Program

GLOSSARY

Absolute value function family

Informally, the family of functions whose graphs have the V-shape of the graph of the absolute value function defined by the equation $y = |x|$.

Example: The function defined by the equation $y = |2x + 1|$, whose graph is shown here, is considered a member of the absolute value function family.

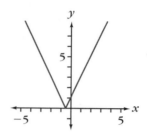

Acceleration

See **velocity.**

Amplitude

See **periodic function.**

Asymptote

Informally, a line or curve to which the graph of an equation draws closer and closer as the independent variable approaches $+\infty$, approaches $-\infty$, or approaches a specific value.

Example: For the function f defined by the equation $f(x) = \frac{3}{x-1}$, both the horizontal line $y = 0$ (which is the x-axis) and the vertical line $x = 1$ (the dashed line in the diagram) are asymptotes. The graph approaches the line $y = 0$ as x approaches $+\infty$ or $-\infty$, and the graph approaches the line $x = 1$ as x approaches 1.

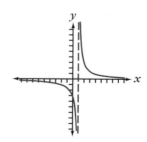

Bias

In sampling, the tendency of a sampling process to overrepresent or underrepresent a portion of the population being sampled. Avoiding bias is an important goal in sampling.

Binomial distribution

A probability distribution describing the result of repeated independent trials of the same event with two possible outcomes. If a particular outcome has probability p for each trial, the binomial distribution states that the probability that this outcome occurs exactly r times out of n trials is $_nC_r \cdot p^r \cdot (1-p)^{n-r}$, where $_nC_r$ is the combinatorial coefficient equal to $\frac{n!}{r!(n-r)!}$.

Example: Suppose a weighted coin has probability .7 of coming up heads. If the coin is flipped 50 times, the probability of getting exactly 30 heads is

$$_{50}C_{30} \cdot (.7)^{30} \cdot (.3)^{20}$$

Central angle

An angle formed at the center of a circle by two radii.

Example: The angle labeled θ is a central angle.

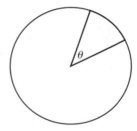

Central limit theorem

A statistical principle stating that when samples of a sufficiently large given size are taken from almost any population, the means of these samples are approximately normally distributed. See *The Central Limit Theorem* in *The Pollster's Dilemma*.

Circular function

Any of several functions defined by placing an angle θ with its vertex at the origin and the initial ray of the angle along the positive x-axis, as shown in the diagram on the next page. If (x, y) is any point different from $(0, 0)$ on the terminal ray of the angle, then the sine, cosine, and tangent of θ are defined by the equations

$$\sin \theta = \frac{y}{r}$$

$$\cos \theta = \frac{x}{r}$$

$$\tan \theta = \frac{y}{x}$$

where $r = \sqrt{x^2 + y^2}$

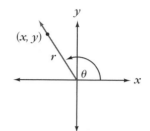

The circular functions also include the secant, cosecant, and cotangent functions, which are defined as other ratios involving x, y, and r. (The circular functions are also called trigonometric functions.)

Commutative

An operation $*$ is commutative if the equation $a * b = b * a$ holds true for all values of a and b. An operation that does not have this property is called *noncommutative*.

Example: The operation of addition is commutative because for any numbers a and b, $a + b = b + a$. The operation of subtraction is noncommutative because, for example, $5 - 9 \neq 9 - 5$.

Completing the square

The process of adding a constant term to a quadratic expression so that the resulting expression is a perfect square.

Example: To complete the square for the expression $x^2 + 6x + 4$, add 5 to get $x^2 + 6x + 9$, which is the perfect square $(x + 3)^2$.

Complex number

A number of the form $a + bi$, where a and b are real numbers and i is the number $\sqrt{-1}$. The number a is the *real part* of $a + bi$, and bi is its *imaginary part*.

Components of velocity

The velocity of an object is sometimes expressed in terms of separate *components of velocity*. For instance, a falling object that is also moving to the side has both a vertical and a horizontal component of velocity.

Example: In the diagram on the next page, if a person is swimming across a river at a speed of 5 feet per second, in the direction shown by the solid line, then the

"toward-shore" component of the swimmer's velocity
is 5 sin 60° feet per second and the "parallel-to-shore"
component of the swimmer's velocity is 5 cos 60° feet per
second. (The "toward-shore" component can also be
expressed as 5 cos 30° feet per second.)

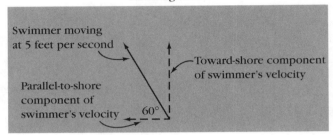

Composition of
functions

An operation on functions in which the output from one
function is used as the input for another. The composition of
two functions *f* and *g* is written *f* ∘ *g*, and this function is
defined by the equation $(f \circ g)(x) = f(g(x))$. A function formed
in this manner is called a *composite function.*

Example: Suppose *f* and *g* are defined by the equations
$f(x) = x^2$ and $g(x) = 2x + 1$. Then *f* ∘ *g* is defined by the
equation $(f \circ g)(x) = (2x + 1)^2$.

Confidence
interval

In a sampling process from a population, an interval around
the sample mean that has a certain likelihood, called the
confidence level, of containing the true mean for the
population. If the interval is symmetric around the sample
mean, then half the length of the interval is called the
margin of error.

Example: The interval of values within one standard deviation
of the sample mean is a *95% confidence interval,* because if
that procedure is followed for many samples, it will contain
the true mean 95% of the time.

Conic section

Any of several two-dimensional figures that can be formed by
the intersection of a plane with a cone. The specific figure
formed generally depends on the angle at which the plane
meets the cone, and is "usually" an ellipse, parabola, or
hyperbola.

Constant function	A function whose value is the same for every input.
Correlation coefficient	A number between -1 and 1, usually labeled r, that measures how well a set of data pairs can be fitted by a linear function. The closer r is to 1, the better the data set can be fitted by a linear function with positive slope; the closer r is to -1, the better the data set can be fitted by a linear equation with negative slope. If r is not close to either 1 or -1, then the data set cannot be approximated well by any linear function.
Cosine	See **circular function.**
Cubic function	A function defined by an equation of the form $y = ax^3 + bx^2 + cx + d$ in which a, b, c, and d are real numbers with a $\neq$ 0.
Dependent variable	See **function.**
Directly proportional	A relationship between two quantities or variables in which one of the variables is a constant multiple of the other variable.
	Example: If an object has been traveling at 20 miles per hour, then the distance it has traveled is directly proportional to the amount of time it has been traveling. This can be seen algebraically as follows: If d represents the distance (in miles) and t represents the time elapsed (in hours), then d and t satisfy the equation $d = 20t$.
Discriminant	See **quadratic formula.**
Domain	The set of values that can be used as inputs for a given function.
	Example: If f is the function defined by the equation $f(x) = \frac{x}{x^2 - 1}$, then the domain of f is the set of all real numbers except 1 and -1.

End behavior

The behavior of a function as the independent variable approaches $+\infty$ or $-\infty$. See *Approaching Infinity* in *The World of Functions*.

Example: For the function defined by the equation $y = 2^x$, the end behavior is that y increases without bound as x approaches $+\infty$ and that y approaches 0 as x approaches $-\infty$.

Exponential function family

The family of functions defined by equations of the form $y = a \cdot b^x$ in which a is a nonzero real number and b is a positive real number other than 1. If $b > 1$, the function is an *exponential growth* function; if $b < 1$, the function is an *exponential decay* function. Functions in this family are characterized by the property that a fixed change in the independent variable always results in the same *percentage* change in the dependent variable.

Fibonacci sequence

The numerical sequence $1, 1, 2, 3, 5, 8, 13, \ldots$, in which the first two terms are both 1 and each succeeding term is the sum of the two preceding terms. (For example, the sixth term, 8, is the sum of the fourth and fifth terms, 3 and 5.) The sequence is often represented using the *recursion equation*

$$a_{n+2} = a_{n+1} + a_n$$

Function

Informally, a relationship in which the value of one variable (the **independent variable**) determines the value of another (the **dependent variable**). In terms of an In-Out table, the independent variable gives the input and the dependent variable gives the output. In terms of a graph, the independent variable is generally shown on the horizontal axis and the dependent variable on the vertical axis.

Formally, a function is a set of number pairs for which two different pairs cannot have the same first coordinate.

Example: The equation $y = x^2$ expresses y as a function of x.

Greatest integer function	See **step function.**

Identity

1. An equation that holds true no matter what numbers are substituted for the variables (as long as the expressions on both sides of the equation make sense).

   Example: The equation $(a + b)^2 = a^2 + 2ab + b^2$ is an identity, because this equation holds true for all real numbers a and b.

2. For a given operation, an *identity* (or an *identity element*) for that operation is an element which, when combined with any element using the given operation, yields that second element as the result.

   Examples: The number 0 is the identity for addition because $x + 0$ and $0 + x$ are both equal to x for any number x. Similarly, the number 1 is the identity for multiplication.

Identity function

The function on a given domain whose output is equal to its input. This function is the identity for the operation of composition of functions and can be represented by the equation $f(x) = x$.

Independent events

Two (or more) events are independent if the outcome of one does not affect the outcome of the other.

Independent variable

See **function.**

Inverse

If an operation has an **identity,** then an *inverse* for a given element (under that operation) is an element which, when combined with the given one, yields the identity as the result.

Examples: The number -7 is the inverse for 7 for the operation of addition because both $7 + (-7)$ and $(-7) + 7$ are equal to 0, which is the identity for addition. Similarly, the number $\frac{1}{5}$ is the inverse for 5 for the operation of multiplication.

*Inverse of
a function*

Informally, the inverse of a function f is the function that "undoes" f. Formally, the inverse of f is its inverse with regard to the operation of composition. That is, the inverse of f is the function g for which both $f \circ g$ and $g \circ f$ are the appropriate identity functions, with $(f \circ g)(x) = x$ and $(g \circ f)(x) = x$. The inverse of f is sometimes represented by f^{-1}.

Example: If f is the function defined by the equation $f(x) = 3x + 2$, then the inverse of f is the function g given by the equation $g(x) = \frac{x-2}{3}$. The fact that g "undoes" f is illustrated by the fact that $f(5) = 17$ and $g(17) = 5$. In terms of composition, we have $(f \circ g)(17) = 17$ and $(g \circ f)(5) = 5$.

*Inversely
proportional*

A relationship between two quantities or variables in which one of the variables is obtained by dividing some constant by the other variable.

Example: If the length and width of a rectangle are to be chosen so that the rectangle has an area of 30 square inches, then the length will be inversely proportional to the width. This can be shown algebraically as follows: If L represents the length and W represents the width (both in inches), then L and W satisfy the equation $LW = 30$, so $L = \frac{30}{W}$.

Isometry

A geometrical transformation T with the property that for any pair of points A and B, the distance between $T(A)$ and $T(B)$ is equal to the distance between A and B. Isometries do not change the size or shape of geometric figures, and include **translations, rotations,** and **reflections.**

*Least squares
method*

A method of determining a function from a given family that best fits a set of data. For a finite set of data points (x_1, y_1), $(x_2, y_2), \ldots, (x_n, y_n)$, the least-squares method seeks the function f (from the family) that minimizes the value of the expression

$$\sum_{i=1}^{n} [y_i - f(x_i)]^2$$

in which $y_i - f(x_i)$ represents the vertical distance between the graph of f and the data point (x_i, y_i).

Linear equation	For one variable, an equation of the form $ax + b = 0$, in which a and b are real numbers with $a \neq 0$. For n variables $x_1, x_2, \ldots, x_n$, an equation of the form $a_1x_1 + a_2x_2 + \ldots + a_nx_n + b = 0$.
Linear function	For a function with one input variable, a function defined by an equation of the form $y = ax + b$ in which a and b are real numbers. (The special case of a constant function, in which $a = 0$, is sometimes excluded from the family of linear functions.) The definition is similar for functions with more than one input variable.
Linear regression	A process for obtaining the linear function that best fits a set of data. The equation for this function is the *regression equation* and its graph is the *regression line*.
Logarithmic function family	Informally, the family of functions defined by equations of the form $y = a + b \log x$ (or of the form $y = a + b \ln x$), where a and b are real numbers with $b \neq 0$.
Loop	In programming, a set of instructions that specifies the repeated execution of a given set of steps.
Margin of error	See **confidence interval.**
Mean of a discrete probability distribution	If a probability distribution has possible outcomes $x_1, x_2, \ldots, x_n$, and the outcome x_i has probability $P(x_i)$, then the mean of the distribution is given by the expression $$\sum_{i=1}^{n} P(x_i) \cdot x_i$$ The mean of the distribution is numerically equal to the expected value of the event that the distribution describes. See *Mean and Standard Deviation for Probability Distributions* in *The Pollster's Dilemma*.
Nested loop	A programming loop that occurs within the body of another loop.

| Normal curve | The graph that represents a normal distribution. If the normal distribution has mean μ and standard deviation σ, then the equation of its graph is |

$$y = \left(\frac{1}{\sigma\sqrt{2\pi}}\right) \cdot e^{-\frac{1}{2}\left(\frac{x-\mu}{\sigma}\right)^2}$$

Example: If $\mu = 0$ and $\sigma = 1$, the equation simplifies to

$$y = \left(\frac{1}{\sqrt{2\pi}}\right) \cdot e^{-\frac{1}{2}x^2}$$

This special case is called the *standard normal curve*. The diagram here shows the graph of this equation.

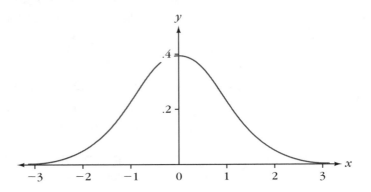

| Parabola | The general shape for the graph of a quadratic function. Also see **conic section.** |

| Parameter | Often, a variable whose value is used to specify a particular member of a family of functions. (The term *parameter* has other meanings as well.) |

Example: The family of quadratic functions consists of all functions defined by equations of the form $y = ax^2 + bx + c$, where a, b, and c are real numbers with a $\neq$ 0. The variables a, b, and c are parameters whose numerical values specify a particular quadratic function. The set of quadratic functions is called a *three-parameter family.*

| Period | See **periodic function.** |

Periodic function   Informally, a function whose values repeat after a specific interval. Specifically, a function f is periodic if there is a positive number a such that $f(x + a) = f(x)$ for all values of x. The smallest positive value for a is called the **period** of f.

If a periodic function has a maximum and a minimum value, then half the difference between these values is the **amplitude** of the function.

Example: The function f defined by the equation $f(x) = 3 \sin (2x) + 5$ is a periodic function with period 180°, because $f(x + 180°) = f(x)$ for all values of x. The graph shown here for this function illustrates its periodic behavior. The maximum value for f is 8 [for instance, $f(45°) = 8$] and the minimum value is 2 [for instance, $f(135°) = 2$], so the amplitude of f is $\frac{1}{2}(8 - 2) = 3$.

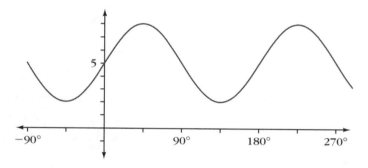

Polar coordinates   A system in which a point in the plane is identified by means of a pair of coordinates (r, θ) where r is the distance from the origin to the point and θ is the angle between the positive x-axis and the ray from the origin through the point, measured counterclockwise. See *A Polar Summary* in *High Dive*.

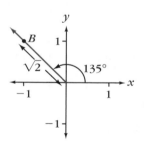

Example: The point B in the diagram, whose rectangular coordinates are $(-1, 1)$, has polar coordinates $(\sqrt{2}, 135°)$ because the distance from $(0, 0)$ to B is $\sqrt{2}$ and the angle from the positive x-axis to the ray through B is 135°. [*Note:* The point B has other polar coordinate representations, such as $(-\sqrt{2}, 315°)$ or $(\sqrt{2}, 495°)$.]

Polynomial function

A function defined by an equation of the form $y = a_n x^n + a_{n-1} x^{n-1} + \ldots + a_1 x + a_0$, in which a_n, $a_{n-1}, \ldots, a_1$, and a_0 are real numbers. If $a_n \neq 0$, the polynomial has *degree n*. The family of polynomial functions includes **constant functions** (degree 0), **linear functions** (degree 1), **quadratic functions** (degree 2), and **cubic functions** (degree 3), as well as functions of higher degree.

Example: The function defined by the equation $y = 3x^4 - 2x^2 + 5x - 1$ is a polynomial function of degree 4.

Power function

Generally, a function defined by an equation of the form $y = ax^b$ in which a and b are real numbers. (In some contexts, restrictions are imposed on b, such as requiring that b be an integer.)

Example: The function defined by the equation $y = 5x^{\frac{1}{2}}$ is a power function.

Principal value

A term sometimes used in the definitions of the inverse trigonometric functions to identify a specific number whose sine, cosine, or tangent is a given value.

Example: The equation $\sin x = 0.5$ has infinitely many solutions, but the solution $x = 30°$ is selected as the principal value, so that $\sin^{-1}(0.5)$ is defined to be $30°$.

Probability distribution

A set of values giving the probability for each possible outcome for an event.

Example: If a fair coin is flipped twice and we count the number of heads, the probability distribution is $P(2 \text{ heads}) = \frac{1}{4}$, $P(1 \text{ head}) = \frac{1}{2}$, $P(0 \text{ heads}) = \frac{1}{4}$.

Projection

A process for representing a three-dimensional object by means of a two-dimensional figure, or any representation of a figure by a lower-dimensional figure.

Pythagorean identity	Any of several trigonometric identities based on the Pythagorean theorem.

Example: The equation $\sin^2 x + \cos^2 x = 1$, which holds true for all values of x, is a Pythagorean identity.

Quadratic equation

For one variable, an equation of the form $ax^2 + bx + c = 0$ in which a, b, and c are real numbers with $a \neq 0$.

Quadratic formula

A formula for finding the solutions to a quadratic equation in terms of the coefficients. Specifically, for the quadratic equation $ax^2 + bx + c = 0$ (where a, b, and c are real numbers with $a \neq 0$), the solutions are given by the quadratic formula expression $\frac{-b \pm \sqrt{b^2 - 4ac}}{2a}$. The expression $b^2 - 4ac$ in this formula is called the **discriminant** of the equation.

Quadratic function

A function defined by an equation of the form $y = ax^2 + bx + c$ in which a, b, and c are real numbers with $a \neq 0$.

Radian

The measure of a central angle of a circle that intercepts a portion of the circumference whose length is equal to the radius of the circle. A radian is approximately equal to $57°$.

Example: In this diagram, the length of the arc from A to B is equal to the length of the radius $\overline{OA}$ of the circle, so angle AOB measures one radian.

Random

A term used in probability to indicate that any of several events is equally likely or, more generally, that an event is selected from a set of events according to a precisely described probability distribution.

Range

The set of values that can occur as outputs for a given function.

Example: If f is the function defined by the equation $f(x) = x^2$, then the range of f is the set of all nonnegative real numbers.

Rational function   A function that can be expressed as the quotient of two polynomial functions.

Examples: The function defined by the equation $y = \frac{3}{x}$ is a rational function, as is the function defined by the equation $y = \frac{x^2 + 3x - 7}{2x^3 + 4x + 1}$. A polynomial function is a special type of rational function. For instance, the function defined by the equation $y = x^3 - 2x^2 + 3$ can be expressed as a quotient of polynomials by writing the equation as $y = \frac{x^3 - 2x^2 + 3}{1}$.

Rectangular coordinates   A system in which a point is identified by coordinates that give its position in relation to each of the mutually perpendicular coordinate axes. In the plane, these axes are usually called the x-axis (horizontal) and y-axis (vertical). The horizontal coordinate is given first.

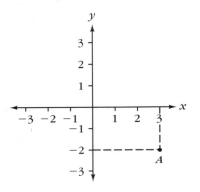

Example: In this diagram, point A has coordinates $(3, -2)$ because it corresponds to the number 3 on the x-axis and the number -2 on the y-axis.

Recursion   A process for defining a sequence of numbers by specifying how to obtain each term from the preceding term(s). An equation specifying how each term is defined in this way is called a *recursion equation*.

Example: The sequence $1, 3, 5, 7, \ldots$ (the sequence of positive odd integers) can be defined by the recursion equation $a_{n+1} = a_n + 2$ together with the *initial condition* that $a_1 = 1$.

Reflection

A type of isometry in which the output for each point is its mirror image. A reflection is sometimes called a "flip."

Example: The diagram here illustrates a reflection in which the *y*-axis is a *line of reflection*. The triangle in the second quadrant is the reflection of the triangle in the first quadrant. In three dimensions, the role of a line of reflection is replaced by a *plane of reflection*.

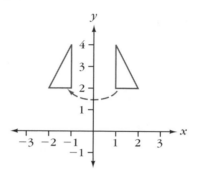

Rotation

A type of isometry in which the output for each point is obtained by rotating the point through an angle of a given size around a given point.

Example: The diagram below illustrates a 90° rotation counterclockwise around the origin. Each point of the lightly shaded triangle is moved to a point that is the same distance from the origin as the original point, but which is 90° counterclockwise (with respect to the origin) from the original point. The diagram shows the "paths" of two of the vertices of the lightly shaded triangle.

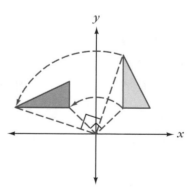

Sample mean

The mean for a sample taken from a population. The sample mean is often used to estimate the **true mean** for the population.

Sample proportion	In a sample from a population, the fraction of sampled items that represent a specific outcome. The sample proportion is often used to estimate the **true proportion.**
	Example: If an election poll of 500 voters shows 285 voters favoring a certain candidate, the sample proportion for the poll is $\frac{285}{500}$, which may be expressed as .57 or as 57%.
Sampling	A process of selecting members of a population and studying their characteristics in order to estimate or predict certain characteristics of the entire population. The selected members of the population comprise the *sample.*
	In selecting the sample at random from a population, we sometimes consider a member of the population ineligible for further selection once it has already been selected. This is called *sampling without replacement.* If members of the population are eligible for repeated selection, this is called *sampling with replacement.*
Sine	See **circular function.**
Sine family of functions	The family of functions whose graphs have the same shape as the graph of the sine function. These functions can be written in the form $y = a \sin(bx + c) + d$, where a, b, c, and d are real numbers with a and b not equal to 0. This family includes the cosine function, because $\cos x = \sin\left(x + \frac{\pi}{2}\right)$.
Standard deviation of a discrete probability distribution	If a probability distribution has possible outcomes $x_1, x_2, \ldots, x_n$, and the outcome x_i has probability $P(x_i)$, then the standard deviation of the distribution is given by the expression

$$\sqrt{\sum_{i=1}^{n} P(x_i) \cdot (x_i - \mu)^2}$$

where μ is the mean of the distribution. See *Mean and Standard Deviation for Probability Distributions* in *The Pollster's Dilemma.*

Step function	Informally, a function whose graph consists of horizontal line segments.

Example: The **greatest integer function,** written $[x]$, is defined by the condition that $[x]$ is the largest integer N such that $N \leq x$. The diagram here shows the graph of this step function.

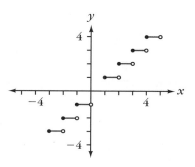

Tangent	See **circular function.**
Transformation	1. A *transformation of a function* is any of certain changes to a function that shift the graph vertically or horizontally or that stretch or shrink the graph vertically or horizontally.
	2. A *geometrical transformation* is a function whose domain is the set of the points in the plane or the set of points in 3-space, in which the image of each point is another point (possibly the same point). An **isometry** is a special type of geometrical transformation.
Translation	A type of isometry in which the output for each point is obtained by moving the point a fixed amount in each of the coordinate directions. A translation is sometimes called a "slide."

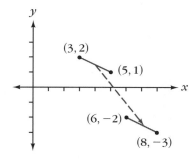

Example: In the diagram here, the line segment connecting $(3, 2)$ and $(5, 1)$ is being translated 3 units to the right and 4 units down, so that its image under the translation is the line segment connecting $(6, -2)$ and $(8, -3)$.

True mean	The mean for a population. See also **sample mean.**
True proportion	In a population, the fraction of items that represent a specific outcome. See also **sample proportion.**
	Example: If 600 out of 1000 balls in an urn are red, then the true proportion of red balls is $\frac{600}{1000}$, which may be expressed as .6 or as 60%.
Unit circle	A circle of radius 1, especially the circle in the coordinate plane with radius 1 and center at the origin.
Variance	The square of standard deviation. For a finite set of data $x_1, x_2, x_3, \ldots, x_n$ with mean μ, the variance is given by the expression

$$\frac{1}{n} \sum_{i=1}^{n} (x_i - \mu)^2$$

Velocity	A combination of the speed and direction of a moving object. If an object is moving vertically, we generally consider upward motion as having a positive velocity and downward motion as having a negative velocity. The rate at which the velocity of an object is changing is the object's **acceleration.** A positive acceleration corresponds to an increase in velocity.

Teacher Book Interior Photography

ix Lynne Alper; **xi** FPG International; **xii** Black Star; **5** Lynne Alper; **20** Lynne Alper; **29** San Lorenzo Valley High, CA; Sandie Gilliam, Lynne Alper; **138** Jim Short; **154** Brookline High School, MA; Terry Nowak, Lynne Alper; **188** Brookline High School, MA; Terry Nowak, Lynne Alper; **218** Capuchino High School, CA; Peter Jonnard, Hillary Turner and Richard Wheeler; **252** San Lorenzo Valley High, CA; Sandie Gilliam, Lynne Alper; **260** Capuchino High School, CA; Peter Jonnard, Hillary Turner and Richard Wheeler.

Student Book Interior Photography

3 Oxnard High School, CA; Jerry Neidenbach; **5** Tony Stone Images; **7** Hillary Turner; **14** Capuchino High School, CA; Chicha Lynch, Hillary Turner, Richard Wheeler; **20** Tony Stone Images; **23** PhotoEdit; **27** PhotoDisc; **30** Capuchino High School, CA; Chicha Lynch, Hillary Turner, Richard Wheeler; **34** SuperStock, Inc.; **37** FPG International; **38** Capuchino High School, CA; Peter Jonnard, Hillary Turner, Richard Wheeler; **41** Tony Stone Images; **43** The Image Bank; **44** FPG International; **45** FPG International; **48** Capuchino High School, CA; Peter Jonnard, Hillary Turner, Richard Wheeler; **52** PhotoDisc; **53** SuperStock, Inc.; **54** Capuchino High School, CA; Chicha Lynch, Hillary Turner, Richard Wheeler; **55** Hillary Turner; **62** Stock Boston; **68** Capuchino High School, CA; Chicha Lynch, Hillary Turner, Richard Wheeler; **70** Tony Stone Images; **72** Tony Stone Images; **78** Capuchino High School, CA; Peter Jonnard, Hillary Turner, Richard Wheeler; **77** The Image Bank; **79** The Image Bank; **80** PhotoEdit; **82** SuperStock, Inc.; **85** Leo de Wys, Inc.; Tony Stone Images; **87** FPG International; **91** Capuchino High School, CA; Peter Jonnard, Hillary Turner, Richard Wheeler; **93** SuperStock, Inc.; **98** Corbis/Bettmann; **99** FPG International; **102** Leo de Wys, Inc.; **104** SuperStock, Inc.; **110** Leo de Wys, Inc.; **113** Capuchino High School, CA; Peter Jonnard, Hillary Turner, Richard Wheeler; **114** Hillary Turner; **118** Hillary Turner; **119** Hillary Turner; **121** Aptos High School, CA; Anthony Pepperdine; **124** Hillary Turner; **129** The Image Bank; **130** SuperStock, Inc; **131** SuperStock, Inc; **133** Brookline High School, MA; Terry Nowak, Lynne Alper; **140** SuperStock, Inc.; **142** Capuchino High School, CA; Peter Jonnard, Hillary Turner, Richard Wheeler; **143** Stock Boston; **149** Tony Stone Images; **150** Stock Boston; **153** Stock Boston; **159** Shasta High School, CA; Dave Robathan; **161** Leo de Wys, Inc.; **163** FPG International; **176** SuperStock, Inc.; **182** PhotoDisc; **184** Capuchino High School, CA; Chicha Lynch, Hillary Turner and Richard Wheeler; **187** SuperStock, Inc.; **188** The Image Bank; **190** Foothill High School, CA; Cheryl Dozier; **208** Animals, Animals; **210** The Image Works; **215** Capuchino High School, CA; Dean Orfanedes; **216** FPG International; **222** Hillary Turner; **233** Hillary Turner, Richard Wheeler; **236** Corbis/Bettmann; **241** Hillary Turner, Richard Wheeler; **242** Brookline High School, MA; Terry Nowak, Lynne Alper; **243** Hillary Turner; **257** Foothill High School, CA; Cheryl Dozier; **258** Tony Stone Images; **259** Leo de Wys, Inc.; **262** Palm Press/©Harold E. Edgerton; **265** Capuchino High School, CA; Dean Orfanedes, Hillary Turner, Richard Wheeler; **280** FPG International; **281** FPG International; **282** Capuchino High School, CA; Dean Orfanedes,

Hillary Turner, Richard Wheeler; **284** Hillary Turner; **291** Capuchino High School, CA; Chicha Lynch; **292** Corbis; **293** PhotoEdit; **300** San Lorenzo Valley High School, CA; Sandie Gilliam, Lynne Alper; **307** PhotoEdit; **308** Capuchino High School, CA; Peter Jonnard, Hillary Turner, Richard Wheeler; **310** Hillary Turner; **312** Tony Stone Images; **314** Hillary Turner, Richard Wheeler; **315** San Lorenzo Valley High School, CA; Sandie Gilliam, Lynne Alper; **320** SuperStock, Inc.; **335** San Lorenzo Valley High School, CA; Sandie Gilliam, Lynne Alper; **341** Foothill High School, CA; Madeline Rippe, Cheryl Dozier; **348** Leo de Wys, Inc.; **357** The Image Bank; **359** Comstock; **361** PhotoDisc; **365** Capuchino High School, CA; Chicha Lynch, Hillary Turner, Richard Wheeler; **373** FPG International; **377** Brookline High School, MA; Terry Nowak; **386** Capuchino High School, CA; Chicha Lynch, Hillary Turner, Richard Wheeler; **395** Hillary Turner; **400** PhotoEdit; **409** Santa Cruz High School, CA; George Martinez, Lynne Alper; **414** SuperStock, Inc.; **415** SuperStock, Inc.; **416** FPG International; **417** The Image Bank; **420** SuperStock, Inc.; **421** SuperStock, Inc.; **423** Patrick Henry High School, MN; Jane Kostik; **429** The Image Bank; **430** Black Star; **431** SuperStock, Inc.; **434** The Image Works; **436** Foothill High School, CA; Cheryl Dozier; **445** FPG International; **447** The Image Bank.

Cover Photography

High Dive Corbis and Comstock; *Know How* Tony Stone Images, Inc.; *As the Cube Turns* Hillary Turner and Rick Helf; *The World of Functions* Hillary Turner and Richard Wheeler; *The Pollster's Dilemma* Corbis; *Back cover* Brookline High School, CA; IMP teacher Terry Nowak.

Front Cover Students

The World of Functions first row: Hilda Chavez, Mary Truong, Tom Hitchner, Enrique Gonzales. Second Row: Kermit Bayless, Jr., Kei Takeda, Ryan Alexander-Tanner, Rena Davis.

Interactive Mathematics Program
Year 4
Comment Form

I M P

Please take a moment to provide us with feedback about IMP. If you have comments or suggestions about Year 4, we'd like to read them. Once you've filled out this form, all you have to do is fold it and drop it in the mail. We'll pay the postage. Thank you!

Your Name _____

School _____

School Address _____

City/State/Zip _____

Phone _____

1. List any comments about the IMP *Year 4* student text.

_____

_____

_____

_____

_____

_____

2. List any comments about the teacher's guide for _____.

_____

_____

_____

_____

3. Do you have any other comments about IMP *Year 4* or any suggestions for improving the student text or teacher material?

_____

_____

_____

_____

Thank you for taking the time to fill out this comment form.

Please return completed forms to:

   Editorial—IMP, Key Curriculum Press, Box 2304, Berkeley, CA 94702.

   You can fold this form as shown on the back and it becomes a postage-paid self-mailer!

Fold carefully along this line.

- -

BUSINESS REPLY MAIL

FIRST CLASS MAIL   PERMIT NO. 151   BERKELEY , CA

POSTAGE WILL BE PAID BY ADDRESSEE

KEY CURRICULUM PRESS
Innovators in Mathematics Education

P.O. Box 2304
Berkeley, CA 94702-9983
Attn: Editorial—IMP

- -

Fold carefully along this line.